Energy and the Environment

Energy and the Environment

The Linacre Lectures 1991–2

Edited by

BRYAN CARTLEDGE

Principal of Linacre College
University of Oxford

Oxford New York Tokyo
OXFORD UNIVERSITY PRESS
1993

Oxford University Press, Walton Street, Oxford OX2 6DP

Oxford New York Toronto
Delhi Bombay Calcutta Madras Karachi
Kuala Lumpur Singapore Hong Kong Tokyo
Nairobi Dar es Salaam Cape Town
Melbourne Auckland Madrid

and associated companies in
Berlin Ibadan

Oxford is a trade mark of Oxford University Press

Published in the United States
by Oxford University Press Inc., New York

A catalogue record for this book is available from the British Library

Library of Congress Cataloging in Publication Data

Energy and the environment / edited by Bryan Cartledge.
(The Linacre lectures; 1991–92)
Includes index.
1. Energy development—Environmental aspects.   2. Sustainable
development.   I. Cartledge, Bryan, Sir.   II. Series: Linacre
lecture; 1991–92.
TD195.E49E512   1993   333.79´14—dc20   92-43543
ISBN 0-19-858413-X
ISBN 0-19-858419-9 (pbk.)

Typeset by Joshua Associates Ltd., Oxford
Printed in Great Britain by
Biddles Ltd., Guildford and King's Lynn

Acknowledgements

This second series of Linacre Lectures, like the first, was made possible by the generosity and environmental commitment of the Racal Electronics Group: Racal's partnership in this venture with Linacre College has continued as smoothly and fruitfully as it began, and for this both the College and Oxford University are sincerely grateful.

I must also express, once again, my warm thanks to Frances Morphy for shouldering the main burden of editing our lecturers' texts and for helping me to meet the exacting standards of our publishers, the Oxford University Press. It is, finally, a pleasure to record my gratitude to Linacre's College Secretary, Jane Edwards, whose enthusiasm and efficiency in administering the lecture series ensured that it was (usually) all right on the night.

Linacre College, Oxford   B.G.C.
September 1992

Contents

Authors

Sir Bryan Cartledge
Principal, Linacre College, Oxford

Professor Sir Hermann Bondi, FRS
Emeritus Professor of Mathematics, King's College London, and formerly Master of Churchill College, Cambridge

Jonathon Porritt
Freelance writer and broadcaster, former Director of Friends of the Earth, and former Chair of the Green Party

Dr Peter Chester, FEng
Former Executive Director for Technology and Environment, National Power plc, currently Chairman of National Wind Power Limited

Sir Denis Rooke, FRS, FEng
Chancellor, Loughborough University of Technology, and formerly Chairman of British Gas plc

Dr John Rae
Chief Executive, Atomic Energy Authority Environment and Energy

Dr Peter Hardi
Director, Regional Environmental Centre for Central and Eastern Europe

Dr Zhores A. Medvedev
Formerly of the National Institute for Medical Research, London, and, until his exile from the Soviet Union, Senior Scientist, Institute of Physiology and Biochemistry of Farm Animals, Borovsk

Lord Marshall of Goring, FRS

Chairman, World Association of Nuclear Operators, formerly Chairman of the Central Electricity Generating Board and of the United Kingdom Atomic Energy Authority

Introduction

Bryan Cartledge

Few areas of public policy confront the governments of the world with sharper or more complex dilemmas than that of energy generation. This second series of Linacre Lectures exposes and defines the dilemmas, principally in the environmental context; but, as several of the lecturers acknowledged, few governments can give absolute priority to minimizing the adverse environmental impact of energy generation. A few countries, blessed in this respect by geography, are able to generate all or most of their electricity from the renewable and (usually) environmentally benign resource of hydropower: in Europe, Norway and Iceland and to a lesser extent Austria and Switzerland are in this fortunate position. As Peter Hardi shows, however, even hydro-electric power can be highly controversial in the environmental context: the Nagymáros dam project in Hungary, which would have deformed the outstandingly beautiful Danube Bend, has been suspended in deference to sustained public protest and to Hardi's own report on the environmental impact of the project.

In determining their energy policies, or the activities which private corporations may be allowed to pursue, most governments in the developed world are obliged to balance three primary factors which are frequently in mutual conflict: economic benefit, public acceptability, and environmental impact. In the developing world, governments are typically concerned with only the first of these criteria and a very few of them, the rare democracies of the Third World, with the second: none of them can afford to devote much money or effort to the environmental dimension. Hermann Bondi highlights these differing priorities. All the lectures, however, point up the dilemmas and the fact that even within the dominant criterion of economic benefit there are built-in contradictions which each government has to resolve in the light of its particular circumstances and constraints. To major coal producers such as China, India, and Poland, the relatively high efficiency and thus economic attraction of gas-fired turbines is irrelevant: relatively inefficient coal-fired power stations, with high emissions of carbon dioxide (CO_2) and sulphur dioxide (SO_2), are the only rational option.

The main thrust of these lectures, however, is directed towards those countries, including Britain, which are well into the third generation of technology and therefore in a position to make choices in energy policy and to arrive at an optimal balance between the economic, political, and environmental criteria. In terms of energy efficiency—typically less than 50 per cent— and of environmental impact (despite the new advances in coal technology which Peter Chester describes), coal and oil score low marks. The scope for reducing CO_2 emissions from coal-fired power stations is very limited, while the prospects for removing CO_2 by extraction prior to emission depend on resolving the formidable technical and financial problems of CO_2 storage, which are at least as complex as those of storing nuclear waste. The only significant advantage of coal is that there is plenty of it. The same cannot be said for oil, which at current depletion rates is likely to cease to be an economically viable option well before the middle of the next century.

Oil, like natural gas, is saddled with a further disadvantage of which governments have to—or at least should—take account. If optimists cherished the illusion that the end of the Cold War would usher in a new era of international harmony and security, the events of 1991–2 should have disabused them. Aggression in the Gulf, near genocide in what was Yugoslavia, and a burning necklace of civil wars on the periphery of the former Soviet Union have shown that the relative certainties of a bipolar world have given way to a much less predictable, more anarchic political climate; regional conflict and terrorism will be endemic until nations are galvanized into the creation of an effective inter-national security system. Meanwhile, energy resources which depend on pipelines, tankers, and offshore rigs will be vulnerable not only, as hitherto, to accident but also to a growing risk of sabotage. Both are environmentally damaging. More importantly, as John Rae points out, the hazards of war and terror should inhibit governments from allowing their economies to become too dependent on the most vulnerable sources of energy.

Apart from vulnerability—and, like oil, the probability of virtual exhaustion by 2050—natural gas, as both Denis Rooke and Peter Chester demonstrate in their lectures, has a great deal going for it. Particularly when used to drive combined cycle gas turbines (CCGTs), it is relatively efficient and, again relatively, benign in its environmental impact. This makes it a more attractive option—for those in a position to choose—than either coal or oil. Technological advance in coal gasification may enable the economic and environmental advantages of gas to be combined with the much greater life expectancy of the coal resource.

The alternative, renewable, energy sources which John Rae describes clearly deserve a place in every government's energy portfolio on environmental grounds alone. Continuing research and development, moreover, are likely to make them more efficient and economic. The renewables—wind power, fuel wood, solar power, landfill gas, tidal power, wave power, and the rest—are nevertheless unlikely to constitute more than a useful subsidiary element in a country's energy strategy. Although wind power, for example, could in theory meet all Western Europe's energy needs, it simply is not realistic, as John Rae allows, to contemplate covering the entire continent with the forest of windmills which this goal would require. Wood and straw fuels, though easily renewable, are by no means pollutant free. Equally, it would be wrong to confine the environmental criterion to air quality alone: a tidal barrage could cause significant ecological damage, and the generation of wave power on a large scale could impair the beauty of coastlines to an extent which the public might be unwilling to accept. Renewability is, literally, a priceless characteristic, but the alternative renewables cannot by themselves provide the optimal answer to energy supply which the governments of developed and environmentally aware countries are seeking.

I have less reason than most to be starry eyed about nuclear power. When, at 01.24 on Saturday, 26 April 1986, Unit 4 of the Chernobyl AES (Atomic Energy Station) exploded and precipitated the world's worst civil nuclear catastrophe, I was serving as British Ambassador in Moscow. Once the fact of the disaster had been established, initially through our Swedish colleagues armed with the evidence of routine atmospheric radiation monitoring near Stockholm, it fell to me and my staff to organize the evacuation of some thirty British students from Kiev. The evacuation was effected in the face of strenuous protests from the Soviet authorities. Late at night on 29 April, twenty-four hours after the first minimal public admission by the Russians that 'a minor accident' had occurred at Chernobyl, I was summoned to the Soviet Foreign Ministry to be told by a triumvirate of senior officials—the Chairman of the State Committee for Atomic Energy, the Minister for Higher Education, and the First Deputy Minister for Foreign Affairs—that our evacuation arrangements were totally unnecessary, a blatant provocation designed to spread panic and anti-Soviet feeling. We persevered: the evacuation was completed on 1 May. The monitoring for radioactivity of foodstuffs purchased in the Moscow markets and the collection of soil and water samples for analysis in London became part of the Embassy's routine for several months.

Just over six months after the disaster, on 17 December 1986, I had the doubtful distinction of being the first Westerner to visit Chernobyl and enter the power station since the explosion, thanks to a visit to the Soviet Union by the then British Secretary of State for Energy, Peter Walker. My impressions from that grey winter's day remain indelible: the barren lifelessness of the exclusion zone around Chernobyl over which our helicopter flew; the steam rising from the cooling lake adjacent to the power station, recalling the pit of Avernus; the eery emptiness of the nearby town of Pripyat', evacuated after the accident, emphasized by the still functioning traffic lights on the deserted streets; the haggard fatigue on the face of the station director, appointed soon after the catastrophe, whose predecessor had already begun a long prison sentence; above all, the difficulty of absorbing the stark fact that an explosion in this small, drab building in the middle of nowhere had threatened the health of every living thing not only in large areas of the Ukraine and Byelorussia but also in much of Scandinavia and of Eastern and Western Europe.

The purpose of this personal digression is simply to establish my credentials for objectivity in the matter of nuclear power. I have seen the worst that it can do. I nevertheless share Lord Marshall's conviction that nuclear power is 'the ultimate source of energy for mankind'. He said to me after his lecture: 'I doubt whether I have made any converts; but the green lobby in 2015 will be lobbying for nuclear power.' I believe he is right. But the case is not an easy one to make. The fears and apprehensions of ordinary people are none the less real for being irrational; governments have to take careful account of them and they are, in any case, more likely to be dispelled—doubtless over a long period—by patient explanation than by impatient dismissal or ridicule. This does not alter the fact that Lord Marshall's debunking of the radiation bogey is supported by all the evidence, provided that the increasingly stringent safety standards in the nuclear power industry are maintained—in fact, they are being steadily enhanced. A catastrophe on the scale and of the nature of Chernobyl could never have happened—and could never happen—outside what used to be the Soviet Empire, not only because the design weaknesses of the RBMK-1000 are unique to that type of reactor, but also because the nature of the human and administrative errors which made the catastrophe possible was a specific product of the Communist system. The accident at Three Mile Island, Pennsylvania, in 1979 showed that Western nuclear reactors are not immune to mechanical malfunction or human error: but the automatic shutdown provision worked and, thanks to the sound construction of the containment vessel, very little hydrogen gas or other toxic gases escaped into the atmosphere—certainly not enough to pose a health threat to the local population. Moreover, the reaction of the federal regulating authority, both in temporarily closing down all reactors of the Three Mile Island type and in instituting enhanced technical and safety standards, was both prompt and comprehensive. The Three Mile Island episode, in fact, provides more grounds for optimism about the future of nuclear power than for apprehension.

Sadly, there can be no guarantee that a repetition of Chernobyl cannot occur *within* the former Soviet Union. Despite the dramatic impact (which Zhores Medvedev describes) of the antinuclear lobby on the ex-Soviet nuclear programme, seven Chernobyl-type (RBMK-1000) reactors which fall short of international safety standards are (in 1992) still in operation, mainly because the Republics concerned (including Ukraine) cannot, in the throes of an energy crisis, afford to dispense with them. Like most of the former Soviet workforce, workers in the nuclear power industry are demoralized by their countries' economic collapse, by rampant inflation, and by irregular and inadequate wage-packets; safety inspectors are leaving nuclear plants in significant numbers in order to seek more lucrative employment. Western financial assistance has been promised to help to rescue the ex-Soviet nuclear power industry but (in August 1992) the sums promised do not begin to measure up to the scale of the problem. If a further accident does occur, in the Commonwealth of Independent States or in Eastern Europe, it will compound the distortion of the public perception of nuclear power which is part of Chernobyl's legacy: but, as in the case of Chernobyl, a further accident will not constitute an argument against nuclear power *per se* any more than a disaster in an unsafe, ill-managed Turkish coal mine constitutes an argument against coal.

In Britain, public concern over nuclear power focuses not only on power stations and their supposed emissions but also, quite naturally, on the problems of storing nuclear waste. Controversy over the local environmental impact of Windscale and of Sellafield in its early phase will doubtless continue for several years yet; it would, in fact, be surprising if the newborn industries of nuclear power and nuclear reprocessing could be shown to be environmentally immaculate from the moment of their inception. The important point is that research and investment—such as the construction of the new Thermal Oxide Reprocessing Plant (THORP) at Sellafield, now nearly complete—

can improve techniques and safety margins to the point at which public concerns will be shown convincingly to be unfounded.

Against this (in my belief) reassuring background, full weight can be given to the fact that in terms of 'greenhouse' and acidic emissions nuclear power is far and away the most acceptable energy source. If Britain's existing nuclear power plants were replaced by gas-fired stations, Britain's CO_2 emissions would increase by 7 million tonnes annually, adding 4.5 per cent to total UK emissions; if our nuclear plants were replaced by coal-fired stations, the annual CO_2 emission would increase by *15.5 million tonnes a year*—an overall increase of *nearly 10 per cent*. An accelerated programme of nuclear reactor construction would therefore be a major and positive contribution to achieving the objectives to which most governments of the world subscribed at the International Conference on Global Climate Change in Rio de Janeiro in 1992, particularly the commitment to stabilize CO_2 emissions at 1990 levels by 2005. As Hermann Bondi points out, a prudent expansion of nuclear power in the developed world can help to offset the inevitable increase in 'greenhouse' emissions resulting from Third World industrialization. Unlike oil and natural gas, uranium will be available in sufficient quantity to fuel even a greatly expanded nuclear power industry for centuries; moreover, nuclear power stations and nuclear waste storage facilities are much easier to protect against violence than pipelines or tankers.

It is the economic argument against nuclear power which deserves to be taken most seriously. Nuclear power stations are and probably always will be more expensive to build than coal, oil, or gas-fired stations. The decommissioning of a nuclear station at the end of its life, moreover, is more complex, much more protracted, and therefore considerably more expensive than the decommissioning of a fossil-fuel-fired station; in addition, fuel reprocessing and the storage of waste are costly operations unique to the nuclear power industry. All these factors add to the unit cost of electricity generated by nuclear power. In Britain, the real cost of 'nuclear' electricity is about twice that of electricity generated by other means: only what is in effect a government subsidy makes it commercially viable.

It is important, however, to distinguish between the British situation and that in other countries whose nuclear industries have a different and more rational history. Not for the first time, Britain is paying for her national preoccupation—particularly marked in the immediate post-imperial era—with being 'first in the field' and 'leading the world' in invention and advanced technology. The crash programme of the 1950s to make Britain the world's pioneer generator of nuclear power has saddled the country with nine Magnox reactors, now nearing the end of their safe life, which are ludicrously inefficient, produce absurd quantities of nuclear waste, and make disproportionate (though lucrative) demands on Sellafield's reprocessing and storage capacities. The same obstinate refusal to adopt or adapt non-British technology produced the programme of advanced gas-cooled reactors (AGRs), now numbering seven, which was launched in the 1970s and only completed in 1989; though less wasteful than the Magnox, the AGR is even less cost efficient and has been a plant-constructor's nightmare from first to last.

The economic case against *British* nuclear generation is therefore powerful. It is not, however, by any means conclusive. In other countries, which have had the good sense to base their nuclear energy programmes on the simpler and by now well-tried pressurized water reactor (PWR), originally designed in the United States, the cost of 'nuclear' electricity is already comparable with that generated by fossil fuels. According to forecasts by the Organization for Economic Co-operation and Development and the International Atomic Energy Agency ('Projected costs of generating electricity from power stations for commissioning in the period 1995–2000', OECD/IAEA 1990), the commissioning of a new, more efficient generation of nuclear reactors during the last five years of this century will give nuclear power a decided edge. In Belgium, the cost of coal-produced electricity is expected to be 1.79 times that of nuclear electricity, in France 1.45 times, in Japan 1.28 times, and in the eastern United States 1.07 times. Coal-fired stations will be further disadvantaged by the capital cost of the modifications which will have to be introduced to reduce CO_2 emissions in accordance with the commitments made at Rio de Janeiro. The economic comparison with the other fossil fuels, too, is likely increasingly to favour nuclear power as oil and gas prices rise in parallel with dwindling reserves. If significant new reserves are discovered anywhere in the world, they are likely to be in remote, hitherto unprospected regions from which the transport of oil or gas will be expensive, thus further loading the economic dice against these fossil fuels.

These factors can and should apply equally in Britain. The government has already been converted, albeit belatedly, to the merits of PWR technology: the Sizewell B station, due for commissioning in 1994, will have a reactor of this type and Nuclear Electric, the state-owned nuclear power generator, is applying for permission to build two more to create a Sizewell C station. Just as Sizewell B will be a significant advance, in economic terms, on the Magnox and AGR stations, so Sizewell C—if it is given the green light—will mark a further advance in cost-effectiveness since it will benefit from the existing infrastructure at Sizewell B. The economic case against British nuclear power will thus become progressively weaker during the next few years and seems certain to collapse before the end of the century. The arguments for expanding the nuclear share of power generation in Britain will then be overwhelming.

I cannot, therefore, agree with Jonathon Porritt's eloquent plea for the diversion of research and development resources away from what he calls Britain's 'prolonged experiment with nuclear power' and towards the development of the renewable energy sources which John Rae describes. On environmental grounds alone, the renewables certainly deserve a higher level of investment and should occupy a more conspicuous place in the British energy portfolio: but that portfolio *must* also contain an expanded, demonstrably safer, and more efficient nuclear power industry. Without it, and given the shrinkage of oil and gas resources, Britain will be driven back to over-dependence on coal and will consequently be incapable of fulfilling the environmental obligations which the British Government rightly undertook at Rio.

POSTSCRIPT

Since this Introduction was written in August, 1992, there have been several developments which have a direct bearing on its conclusions.

In October, 1992 British Coal unexpectedly announced a draconian programme of pit closures which, if fully implemented, will effectively remove the option of making British-mined coal a central pillar of the nation's energy policy. Once decommissioned, coal mines can be re-opened only at vast expense, if at all. Moreover, the heightened preference of the power generators for natural gas—the so-called 'dash for gas'—which, allegedly, made the pit closures an economic necessity will in itself accelerate the depletion of the natural gas resource: according to British Gas, known reserves are now likely to be exhausted in twenty-one years at projected levels of consumption.

In January, 1993 the oil tanker 'Braer' broke up against the southern tip of the Shetland Islands and spilled 84 500 tonnes of oil into Shetland coastal waters—twice as much as the 'Exxon Valdez' had spilled off Alaska in 1989. Experts opined that periodic environmental disasters of this kind are an inevitable concomitant of the large-scale transport of oil by sea.

During the course of a few months, therefore, major new question marks have appeared over the long-term availability of both coal and natural gas; and the environmental disadvantages of oil have been further underlined. These developments should, logically, have strengthened the case for an expansion of Britain's nuclear resource. Instead, the British Government's policy towards the nuclear industry at this crucial juncture seems to be in total disarray. Doubts are being expressed over the desirability and financial feasibility both of proceeding with the Sizewell C project and of commissioning the THORP facility at Sellafield. There is talk of cushioning the Government's own blow to the coal industry by diverting to it part of the nuclear industry's subsidy, the 'nuclear levy', therefore making electricity generated by nuclear power even less economic. If this series of Linacre Lectures helps to underline the urgent need for a coherent and comprehensive energy strategy in Britain, they will have served a further useful purpose.

1

Energy and the environment: the differing problems of the developed and developing regions

Hermann Bondi

Professor Sir Hermann Bondi, KCB, FRS, is one of the great polymaths of British science. After serving in the Admiralty during the Second World War, he was appointed to a lectureship in mathematics at Cambridge University where he also held a Fellowship at Trinity College. In 1954 he became Professor of Mathematics at King's College, London, a post which he held until 1971 and which he still retains in an emeritus capacity. Professor Bondi then began a long and valuable association with Government, first as Chief Scientific Advisor to the Ministry of Defence and then, in 1977, as Chief Scientist at the Department of Energy; in 1980 he was appointed Chairman and Chief Executive of the Natural Environment Research Council. In 1983 Sir Hermann returned to Cambridge on his election as Master of Churchill College, from which post he retired in 1990, while continuing as a Fellow of the College. He holds honorary Doctorates from the Universities of Sussex, Bath, Surrey, York, Southampton, Salford, Birmingham, and St Andrews. In 1988 Sir Hermann was awarded the Gold Medal of the Institute of Mathematics and its Applications. His many publications include works on cosmology, relativity, and astrophysics. He has been, and remains, an indispensable bridge between the world of research and that of policy.

Energy is the key item in our relations with our environment. Energy consumption determines how much and how severely we can affect our environment, and how damaging or healing our interactions with it are. Therefore I will concentrate entirely on energy production, consumption, and use.

The energy output of a sedentary person, like the member of an audience, is about 100 to 120 watts (W). A person engaged in strenuous exercise may put out perhaps fifteen times as much, largely as heat. But one cannot be so active for very long, so that the average energy production of a human being is unlikely to exceed 200 W or thereabouts. This output has to be balanced by food consumption, and this is the primary human energy requirement. In fact, the energy consumption per head (that is, the national energy use divided by the number of inhabitants) is around 8000 W in North America and between 3000 and 5000 W in other industrialized countries. Why is this so much larger than the primary 150–200 W mentioned before?

The enhanced energy consumption began with the start of agriculture. It turns out that the only foodstuffs that can be grown in large quantities need preparation, including cooking, which requires energy. Once the numbers of humans had grown, cooking became an essential. How much energy is needed for cooking? This question leads to an important feature of energy use. What is wanted, in this as in many other cases, is not the energy itself but some objective which cannot be attained without its use. However, the quantity of energy needed for this purpose is generally not intrinsically defined, but is largely dependent on the efficiency of the processes employed. This applies very much to cooking. It may be regarded as paradoxical that in developing countries, where cooking stoves are usually very inefficient, much more energy is needed for food preparation than in industrialized countries, partly because it is far harder to burn firewood efficiently than, say, natural gas. It is worth mentioning that in some situations, such as chemical transformations like making aluminium from its ore, or lifting materials (e.g. raising coal from the bottom of a mine to the surface), there is a physical lower limit to the amount of energy needed, but in practice this is normally exceeded by a large margin. So it will be appreciated that the efficiency of energy use is a very important matter about which I will say more later.

Returning to the use of energy for cooking, this must, in many developing countries, amount to perhaps 100–200 W per head. But while this use of energy is plain, it leads inevitably to the question of what we, in the industrialized world, use so much energy for. We only have to look around to see where energy goes. We live, work, study, and also play in buildings which are made of brick, concrete, steel, and glass. Every one of these materials takes a great deal of energy to make. Moreover, their ingredients have to be obtained and transported to the place of their manufacture, and then the finished material has to be transported to the building site. Transport involves trains, cars, lorries, ships, and aeroplanes. Each of these requires a great deal of energy to make. In use, each of these vehicles needs a lot of energy to make it go. At home, we want bodily comforts such as heating, hot water, and, in many regions, air conditioning. Each of these, particularly perhaps air conditioning, uses a great deal of energy.

Somewhat arbitrarily, one can divide these uses into domestic, industrial, and transport. But please remember that each of these makes and moves things ultimately for us, the final consumers. In this country these three groups each take about the same amount of energy. But transport has been a rapidly growing user. A mere fifteen years ago, it took only a fifth of the total energy consumption.

The industrial world has experienced an enormous growth in prosperity and well-being during my lifetime, especially in the period 1950–74. In that quarter-century, the growth in prosperity was coupled with a huge growth in energy consumption. Accordingly, these two were considered to be closely linked. Indeed, for a long time steel production was regarded as a prime measure of a country's economic success and standing. Steel is of course a very energy-intensive substance to produce, and some of its problems are characteristic of many problems of modern economics. If I again go back to situations I actually experienced myself, in the late forties and early fifties this country suffered a severe shortage of steel. I was puzzled by this, as steel production was about 13 million tons, and at that stage, few ordinary citizens had cars or washing machines or even refrigerators, all largely made of steel. So I wondered and looked to see where all this steel went. It immediately turned out that nearly a quarter of the 13 million tons went into building new steel works, with substantial additional amounts going into building the ships that would bring the iron ore to the steel works and the railway trucks to bring the coke. I felt at the time that it was sheer good luck that we did not all have to throw in what little steel we had just to keep the whole industry going! In retrospect it is clear how unstable such a situation is. Once the expectation of an ever increasing demand for steel disappears, the market for it shrinks sharply. This has been one of the indications of a profound change in the industrialized economies, with important consequences for their energy characteristics. Whereas forty years ago the whole economy appeared to be based on steel production and other such energy-intensive sectors, today we have an entirely different outlook.

There is little alternative to gross domestic product (GDP) as a measure of prosperity, imperfect though it is. Not only does it fail to measure happiness, it also masks certain aspects of modern life. For example, if people, instead of buying their vegetables in shops, grow them themselves, the amount of money changing hands is diminished (and so therefore is the GDP), but they have fresher vegetables and probably enjoy them more. But accepting the GDP as measure of prosperity for lack of a viable alternative, we can now study the interaction between the economy and energy. It is obvious that energy consumption will be less in a recession than in a boom, though this does not make recessions desirable. But it is very worthwhile to see how much energy the various economic activities demand for equal contributions to the GDP. The best figures I know are American, but their applicability is wide. It is almost obvious that steel, heavy chemicals, aluminium, brick, concrete, etc. all need a great deal of energy per unit contribution to the GDP. On the other hand, computers, financial advice, TV and other entertainment, health, education, telecommunications, and the like need relatively little. But their energy needs are not trifling as they all need buildings, heating, and so on. Some of these American figures are quite amusing. It was found that water transport takes more energy per unit earnings than air transport. Everybody knows that, per ton-mile, air transport is far more energy demanding than transport by barge, but the advantages of air freight are so great (for appropriate goods), that the higher costs outweigh and reverse the ton-mile consideration.

Let us now look at the plausible prospects of advanced economies. Which activities would one expect to grow, and which are those where growth looks less likely? The Japanese speak of sunrise and sunset activities, but this is perhaps a little too sharp a distinction. Returning to our lists of energy consumers above, it turns out that most of the probable growth areas involve low energy intensity activities, whereas few of the high energy users can anticipate major growth, except tourism which, because of movement of people by air and hotel building, is quite a heavy energy user in relation to its earnings. So, in general, the basis of future growth and prosperity lies largely outside the 'smoke stack industries' that were the motors of growth forty years or so ago. Incidentally, quite a number of the likely gainers demand high levels of skill which makes me optimistic that education is going to grow. Altogether, it is reasonable to conclude that our potential growth is now largely unhitched from growth in energy consumption.

One can go a little further. Earlier I referred to the importance of energy efficiency. The same results can often be achieved with far less energy than was customary. Through better engineering, better controls, better hardware and software, efficiency can be much improved in many applications.

An encouraging example of advances in this field of energy efficiency is that in this country the amount of primary energy (i.e. counting the energy that has gone into making electricity rather than just the electricity actually consumed) used for heating homes and producing domestic hot water has not changed materially over the last seventy years or so. During this period the population has grown by about 20 per cent, the living space per head has probably doubled, and the level of comfort has improved enormously. Yet all this has been achieved without growth of energy consumption! Of course, it is a matter of taste how one describes this: one can either say that it shows how clever we are now or how stupid people were several decades ago. But the gain in energy efficiency is a fact which has had several desirable by-products, notably the disappearance of London's infamous killer fogs.

There are many other fields where real progress has been

achieved. In aeroplanes there has been major growth in efficiency, in motor cars more modest progress has been largely masked by increases in driving speeds, in power generation and in steel making there have been steady significant advances. One may well regret that advances have not been greater. Here one meets a curious feature of our culture and our system. For some reason, it is easy to attract the best brains and often a good deal of money into new, especially renewable, sources of energy or into improving the efficiency of electricity generation. Public attention, further finance (perhaps modest), and management skills follow readily. By contrast, it is very hard to get management attention, capital, glamour, and brains into questions of improving the efficiency of energy use. This may be because it requires not one big step, but numerous detailed adjustments and replacements to achieve a satisfactory gain, or it may be because in most businesses and households energy expenditure is but a small fraction of overall costs, or there may be other reasons, but it is a fact.

Some ten years ago the Harvard Business Review made a study of what pay-back period for investments was allowed by the boards of American companies before they sanctioned any expenditure. It turned out (to give rough figures) that if the money was needed to increase the quantity produced of a successful product, an eight to ten year period was acceptable, if it was a new product, six years was the maximum allowed, but if the expenditure proposed was to reduce energy costs, even one year was often considered excessive.

Perhaps this is not surprising. In any organization, the scarcest commodity is top management time. How much of its time can be spent deciding on the many small improvements needed to better energy efficiency in comparison with sales, with staffing, with innovation, with other costs? Indeed one might well feel that it is gratifying that in spite of these obstacles a good deal of progress has been made, but with a little extra effort so very much more could have been achieved. The more attention that can be paid to energy efficiency the better. This is the area that offers the greatest rewards in reducing energy consumption, all attainable with attention and a little capital expenditure. I should say here that I belong strictly to the anti-hairshirt school of thought. I certainly believe that energy use could and should be reduced, but not by making ourselves uncomfortable, which is not only unlikely to find public favour but generally yields very little. Much more can be gained by being a little more intelligent. It will need persistent attention and pressure, and then one can look forward to continuing and accelerating progress. Combining this reasoning with the previous analysis of the likely development of the economies of the industrialized countries, I arrive at the conclusion that static or even gently diminishing energy use will be no bar to the growth in prosperity of our kind of country and its ability to give service to its citizens.

Let us now consider the poorest parts of the world. There we have been shocked on so many occasions by the pictures and other evidence of malnutrition, of hunger, indeed of starvation. It is legitimate and right to ask what is the minimum of economic progress needed to avoid such disasters. Clear essentials are food transport and food storage, for good harvests cannot be expected every year in every region. To improve the harvests themselves, irrigation and appropriate fertilizers will be needed. Every one of these necessities requires a great deal of energy. So just meeting that most basic human need, food, a substantial growth in energy consumption will be essential. Add to this the rapid growth of the populations of many of these countries, and the energy needs of the future assume a formidable size in these parts of the globe. Let us now turn from the very poorest of the developing countries to the somewhat more advanced ones. It so happens that of these, I know India a little. There real progress can be observed, progress made possible by a large growth in energy consumption. Twenty years ago, there were periodic famines in various parts of the country. Now such events are virtually unimaginable. I do not think India's great gains have had nearly enough publicity in Britain. How have they been achieved? In the early 1970s, when India had a bad patch, a great deal of aid arrived in the form of food, as grain or rice. It has been estimated that a third of this aid was eaten by rats. Just a lack of concrete storage vessels was responsible for a huge loss, and this problem has been dealt with. Transport has improved greatly. The efficiency of agriculture has become much better. Irrigation is going where it is needed. And so the spectre of starvation has been banished, but a great investment in energy production and use was required. More needs to be done in India and, they, like their neighbour China, rightly have great plans involving a huge increase in energy use. Clean water, still a scarce commodity in India, educational facilities, industries, communications, all cry out for more energy. Looking again at energy consumption per head and remembering that the number of heads in the developing world is unfortunately growing rapidly, it is plain that it is impossible to imagine an acceptable future for the globe in which there is not a much larger consumption of energy than now, essentially because of the legitimate aspirations of the developing world.

Where is all this energy to come from? First, let me return to energy efficiency. In the industrialized countries it is hard to attract enough attention, enough talent, enough capital to this promising field. In developing countries this is vastly more difficult, for they are short of skills of all kinds and of capital. While undoubtedly they should be encouraged and assisted to improve energy efficiency throughout their economies, their advance in this field will be limited and will not obviate their need to use more and more energy. China, already the world's largest producer and user of coal, has plans to increase this by at least 150 per cent. I have heard it said that if the industrial boilers of China could all be run with the average efficiency of British industrial boilers, this increase could be halved, and I have no reason to doubt this statement. But if one is short of trained people of all kinds, the unglamorous grinding away at energy efficiency is not likely to get the attention it deserves.

So it is plain that much more energy will be needed. What are the available sources of energy? The easiest of all to use is oil. It is cheap to transport. We know how to refine it to create tailor-made fuels for different uses. A good device like a diesel engine is not only cheap and efficient, but easy to maintain. When I am approached by people with weird and wonderful ideas for meeting the energy needs of a remote corner of Africa, I always ask them to compare their device in cost and demand for skills with an old diesel with its fuel brought to it on the back of a donkey. Such a machine sets a standard hard to beat. A further advantage of oil is the way it is found in nature: in underground reservoirs under pressure so that when a hole is drilled down from the surface, it just gushes out! However, oil also has disadvantages. Not only is it very unevenly spread round the globe, but the total amount is thought to be limited. Just how soon this limit will seriously affect its availability and price is a matter of debate. The approach to this conjectured limit would presumably take the form that the reservoirs which were easy and cheap to get at would be exhausted first. There would then remain only locations that are hard and expensive to make use of owing to the small size of the reservoir or awkwardness of access. A further disadvantage, shared in different measure with other 'fossil' fuels, is that the products of combustion, particularly but not only carbon dioxide, have undesirable environmental consequences, notably the 'greenhouse effect'.

The second major fuel is coal. The limits to the total quantity of coal resident in the crust of the Earth are thought to be much higher than those for oil. Moreover, it is far more widespread in its occurrence than oil, though the three big countries of China, the former Soviet Union, and the United States have between them about 85 per cent of all the known reserves. But even the remaining 15 per cent are very substantial and should serve the rest of us for several hundred years. But environmentally coal is very undesirable: not only does it produce markedly more carbon dioxide per unit of energy produced, its other combustion products tend to be rather nasty unless an expensive clean-up with constant competent maintenance is undertaken. Moreover, it is more expensive and disagreeable to transport than oil. But its abundance and familiarity are real advantages.

The third major fuel is natural gas. It is essentially methane, which burns very cleanly. It produces less carbon dioxide per unit of energy, and has virtually no other combustion products. However, unburnt methane has a severe greenhouse effect, so much so that the quantity of it escaping from municipal waste tips (where it results from unavoidable processes) contributes, it is said, as much to global warming as the exhaust from motor cars. This shows once again that unglamorous subjects like environmentally acceptable methods for waste handling are of great importance. Transporting gas economically is not easy. Long pipelines are expensive and consume a good deal of energy for pumping. Leaks from the huge pipeline that runs from Siberia to Western Europe can have most undesirable 'greenhouse' consequences. The amount of gas escaping from it was a state secret in the Soviet Union. On the other hand, so much of the energy from burning methane comes from the hydrogen rather than the carbon that one could imagine methods resulting in water vapour being the only gaseous combustion product. Problems of its transportation result in a huge waste of energy: much of the gas that inevitably comes up with the oil in the Middle Eastern oil fields has to be flared off because there is no use for most of it within reasonable distance. How much gas there is in accessible reservoirs is a matter of debate. The majority of experts think that while gas is also limited, the limits are far higher than for oil, so that on the most pessimistic forecasts gas production needs not peak for many decades yet. According to the minority school there is very much more gas available than this.

One way of getting over the transport problem of gas, particularly because intercontinental undersea pipelines are quite beyond our technology, is to cool it so that it becomes liquid. Tankers carrying liquefied natural gas are well established on a modest scale and have not turned out to be as dangerous as it was thought they might be. Yet a growth in this form of traffic by one or two orders of magnitude would be rather frightening.

Nuclear power has its advantages and disadvantages. There is first the major benefit that the unavoidable contamination of the environment is very low. Secondly, the actual quantities of fuel that have to be moved around are very small. Against all this is the need for a highly skilled and very disciplined workforce. The health and environmental penalties for instances of incompetence or indiscipline can be extremely severe as we all know. Such a workforce is hard enough to gather and keep up to the necessary standard in any advanced country, and in most developing countries the formation of such a workforce would deprive the rest of the economy of such people. Then there is the problem, perhaps more political than technical, of nuclear waste disposal. One may regard this and similar objections to nuclear power as psychological rather than 'real', but, as I have often reminded my friends in the field, psychological facts are also facts.

In spite of these difficulties, I would find a universal phasing out of nuclear power a matter of serious concern. With the growing energy needs of the developing world, for the industrialized countries to deprive themselves of a substantial source of energy is, in the long run, bound to reduce the availability and increase the price of other sources (especially the fossil ones) for those least able to pay. The amount of electricity now produced world-wide by nuclear energy would use all the oil output of a very large oil producer such as Iran if generated by oil. I am not an out-and-out advocate of nuclear power, but I regret it very much when, as so often in discussions of this question, the effect on the poorer parts of the globe is not considered thoroughly. In all the vast volume of paper submitted to the Sizewell Inquiry there was, so I understand, hardly any mention of the developing areas. No decision should be arrived at on the basis of a purely parochial discussion.

The whole subject of renewable energies is of great significance. Let me say straight away, that much as I like the term 'renewables', I detest the term 'alternatives'. We need to use everything that is manageable and tolerably cheap, and not begin by regarding some as replacements for others. Given the energy hunger of the world (in industrialized as well as developing areas), all sources of energy will have to be employed if economic growth, particularly in the poorest parts of the globe, is not to be unduly restrained. As a very experienced friend of mine said: 'If we use everything and have a little good luck, we will get by.'

Turning now to the various forms of renewable energy, it is true for them as for the other forms of energy generation that there should be a match between the type of production and the type of use. To give an example, electricity is energy expensive to make, but wonderfully flexible in service. So it may be excellent for heating a school not used as an evening centre and thus requiring to be heated only five days a week in term time for at most ten hours a day. The ability to switch on and off without waste, characteristic of electricity, can there be put to good use. At the opposite end of the scale is a home for old people where heating is required for twenty-four hours a day, seven days a week. For such a place, the flexibility of electricity is of no benefit, and other methods less costly in energy (i.e. without the big energy losses that necessarily occur at power stations) are to be preferred.

One of the best established of the renewables is wind power. Its chief drawback is its intermittent character. This is quite irrelevant for an application such as irrigation, where it does not matter whether the pumping takes place today or tomorrow or the day after. I was greatly impressed to see the Lasithi Plain in eastern Crete, where for hundreds of years thousands of windmills have ensured the irrigation of a fertile and intensively cultivated area. If there are a few calm days nobody loses any sleep over it. By contrast, think of a Scottish island where electricity is produced by a local diesel (with its fuel shipped in barrels since the harbour cannot take tankers) because a cable connection to the mainland would be too expensive. Wind is almost permanent, and so this looks an ideal site for wind power. However, it is not quite so simple. The inhabitants are used to TV which needs a steady voltage. Wind strength varies from minute to minute. So the diesel has to be run in tandem with the wind turbine. As the wind power fluctuates up and down, the diesel power has to go down and up, so that the combined power is steady. This reduces the efficiency of the diesel markedly so that the saving in fuel is likely to be well below that corresponding to the amount of energy produced by wind. Experiments are going on to establish just how all this works out in practice. But there is little doubt that in many locations and for many purposes, wind power will have a useful part to play.

Another type of energy of long standing is biomass in the form of firewood. The interesting point about firewood is that whether it is renewable or not is a matter of administration and control and not of technology. If a type of tree takes twenty years to grow to be useful as firewood, and each year every twentieth tree is harvested, the forest will be a renewable source. If all the trees in an area near a village are felled, not only is this not a renewable source, but the unprotected topsoil is then likely to be leached and removed by heavy rains (only too common in the Tropics). A desert will then be created. This is an extremely serious matter. Humanity can, I am sure, live without oil. It cannot live without topsoil, so that soil erosion is a far more serious threat than depletion of energy sources.

There are many other forms of renewable energy, notably perhaps photovoltaics. This is not the place to list them all. Many of them will find their place in the global energy household of the future, but forms of generation and kinds of use should always be matched.

There are advantages and disadvantages with all systems, and they all do some damage to the environment. But I cannot imagine a world of which I would be a happy citizen in which the growth of the economies of the poorer areas was unduly inhibited by energy constraints. We have to reckon with some damage to the environment and learn to live with it. It would be very wrong first to under-invest in learning about all promising sources of energy, and then to fail to exploit those found useful for the required applications. We will need a lot of different ones for different sites and different applications.

2

Sustainable development: panacea, platitude, or downright deception?

Jonathon Porritt

Jonathon Porritt took a First Class Honours degree in Modern Languages at Magdalen College, Oxford, before teaching for nine years in state schools in London. His work for the Green Party began in 1977, when he stood as one of the Party's few candidates in the local elections, and culminated in his Chairmanship of the Party in 1982–4; in the meantime he had carried the Green standard in two more local elections, two European elections, and two General elections. His political activity, reinforced by his broadcasting and authorship of four books on environmental issues, has made him the best known and most respected spokesman for the environmental lobby. He has probably done more than anybody else to help the 'Greens' to shed their beards-and-sandals image and to demonstrate that the movement has an intellectual cutting edge to be reckoned with. From 1984 to 1990, Jonathon Porritt was Director of Friends of the Earth; he now devotes all his time to writing and broadcasting, as a freelance.

As a concept, sustainable development is to the Environment Movement what 'sound money' was to the Conservative Party during the 1980s: ubiquitous, capable of multiple (and often contradictory) definitions, and in danger of being enfeebled by both these attributes. It seems important at this stage to try and gauge its usefulness in the current debate. To some, it has become a panacea, answering each and every environmental problem of and by itself. For others, it is a convenient linguistic device to put a green gloss on otherwise totally orthodox economic and political ideas.

It is hard to escape the use of this concept today: sustainable development, sustainable growth, sustainable progress, sustainable agriculture, sustainable markets, sustainable yields—about the only thing that is not sustainable is the level of dishonesty with which such an important word is being abused!

The history of the concept is by now well known. First coined by Barbara Ward for her report to the 1972 Stockholm Conference on the Human Environment (of which the Earth Summit in June 1992 was the twentieth anniversary), it lay pretty dormant for the next fifteen years apart from pioneering treatment of it by the International Institute for Environment and Development, whose work in this area remains second to none.

And then, in 1987, the Bruntland Report of the World Commission on Environment and Development, *Our common future*, defined sustainable development as follows: 'meeting the needs of the present without compromising the ability of future generations to meet their own needs' (World Commission on Environment and Development 1987, p. 8).

Actually, 'sustainable' has an even simpler meaning: if something is sustainable, it is capable of being kept going on an indefinite basis—not till the end of the week, or the end of the decade, or even the end of the next century, but *indefinitely*. To refer therefore to 'sustainable extraction techniques of our North Sea oil reserves' is an utter nonsense. Or to suppose that modern farming is even remotely sustainable, as the National Farmers' Union persists in doing, is wilfully misleading. Modern farming depends on a massive input of incontrovertibly *finite* fossil fuels and on a process of mining the soil which will last only as long as the soil is deep.

Indeed the concept of 'sustainable growth' is a contradiction in terms: exponential growth (in either human numbers or volumes of production and consumption) *cannot* be sustained indefinitely off a finite resource base. A growth rate of 3 per cent implies a doubling of production and consumption every twenty-five years. Nobody actually disagrees with that, not even the most manic growthists. But professional Micawbers that they all are, they just go on hoping that something will turn up before their bluff is finally called.

It is worth reminding people that the reason for this extra-ordinary doubling period is that economic growth is *not* linear, it is exponential. It does not grow by a constant amount over a constant time period; it grows at a rate which is proportional to that which is already there, as with the compound rate of interest on a savings account.

In the newly published sequel to *The limits to growth* (Meadows *et al.* 1972), entitled *Beyond the limits* (Meadows *et al.* 1992), the authors provide one of the best demonstrations of exponential economic growth that I've seen:

There is an old Persian legend about a clever courtier who presented a beautiful chessboard to his king and requested that the king give him in exchange one grain of rice for the first square on the board, two grains for the second square, four grains for the third square, and so forth.

The king readily agreed, and ordered rice to be brought from his stores. The fourth square on the chessboard required 8 grains, the tenth square took 512 grains, the fifteenth required 6,384, and the twenty-first square gave the courtier more than a million grains of rice. By the fortieth square a million million rice grains had to be piled up. The payment could never have continued to the sixty-fourth square; it would have taken more rice than there was in the whole world.

Those who talk of 'green growth' (which means almost nothing) or 'sustainable growth' (which is an out-and-out oxymoron) are therefore whistling in the wind. And this, for me, remains the single biggest problem about the Brundtland Report. Impressive though its analysis of unsustainable growth patterns was, it made no effort to redefine growth other than by reiterating a sequence of wishy-washy phrases about 'environment-friendly growth' and so on. This was undoubtedly a quite conscious political choice: a trade-off to bring on board many Third World countries who continue to see any discussion about redefining growth as 'a plot to keep the poor South poor'. But the failure to address that issue has created a serious conceptual vacuum over the last five years (only partially remedied by the work of environmental economists such as David Pearce), which continues to bedevil the whole Earth Summit process.

Green growth may turn out to be less polluting, less wasteful, and more efficient in terms of energy and resources (all of which are highly desirable goals, enthusiastically to be campaigned for), but its adherents still seem to subscribe to one, all-powerful item of economic dogma: that it is only through a permanent process of expansion in production and consumption that it will be possible to meet human needs, improve material standards of living, and ensure that wealth trickles down to the unfortunate billions who have not yet had 'their share of the cake'. We can, it would seem, have our green cake and eat it.

Genuinely sustainable development is indeed possible. Production of timber from forests can be sustained indefinitely, cycle after cycle. The production of food through organic agriculture can be sustained indefinitely. Unlimited energy supplies can be sustained indefinitely from renewable sources such as the sun, wind, waves, and tides. And though it is technically inaccurate to say that even the most commonly occurring raw materials can be mined or extracted on a sustainable basis (in that eventually, however far in the future, they will run out), a combination of re-use, repair, and recycling could slow down the rate of depletion so as to make their supply all but indefinite.

Since the Brundtland Report, there have been many different definitions of sustainability, perhaps the most useful of which has just appeared in a new document published jointly by the IUCN, the World-Wide Fund for Nature, and the United Nations Environment Programme entitled *Caring for the Earth: a strategy for sustainable living* (IUCN/WWF/UNEP 1991). In formulating their own definition of the term sustainability, the report's authors note that (p. 10):

The term [sustainability] has been criticized as ambiguous and open to a wide range of interpretations, many of which are contradictory. The confusion has been caused because 'sustainable development', 'sustainable growth' and 'sustainable use' have been used interchangeably, as if their meanings were the same. They are not. 'Sustainable growth' is a contradiction in terms; nothing physical can grow indefinitely. 'Sustainable use' is applicable only to renewable resources: it means using them at rates within their capacity for renewal.

> *'Sustainable development' is used in this Strategy to mean: improving the quality of human life while living within the carrying capacity of supporting ecosystems.*

How good to see that all-important notion of carrying capacity back in the frame! Since the fierce debate in the early 1970s about *The limits to growth* and other books like *Blueprint for survival* (Goldsmith *et al.* 1972), environmentalists were forced onto the defensive in justifying the somewhat apocalyptic scenarios opened up in them. What should have been a debate about *timing* (i.e. in what kind of time-frame would physical constraints on exponential economic growth really begin to bite, since there should not be any debate at all about the fact that they will, one day, bite) became a futile exchange of mutually exclusive prejudices.

But there are, of course, physical limits to the expansion of every species on Planet Earth, and it is mind-boggling to encounter people who think even now that the human species is somehow an exception to that rule. Any productive system loses the capacity to regenerate itself if it is over-exploited, and the sheer increase in human numbers is already stretching some of those systems to breaking point.

It must therefore remain a source of great concern that so many environmentalists and activists involved in Third World movements and organizations still do not understand the urgency of the population issue. Indeed, many of them have been persuaded, by a weird combination of reactionary Catholicism and ante-diluvian socialism, that population is 'not an issue'.

They are, to be a little blunt about it, fools. Even if the problems of distribution were solved entirely to their satisfaction (for they are absolutely right to point out that a more equitable distribution of the Earth's resources is *just as much* of a problem as population itself), we would still face desperate ecological problems as a consequence of there simply being too many people asking too much of a finite resource base.

As Lester Brown has pointed out (Brown *et al.* 1990), the reason for this escapist response basically arises out of their continuing adherence to outmoded views about population. The theory of 'demographic transition' (based on Frank Notestein's work in the 1940s) identifies three separate stages of population levels: the 'pre-modern stage', where birth and death rates are both high and population grows very slowly; a second stage, where living conditions (especially with regard to diet and public health) improve rapidly, death rates fall, but because birth rates remain high, the population grows rapidly; and a third stage, where economic and social progress reduces the desire for large families, thus restoring the equilibrium between birth and death rates.

Fine so far, but does such a theory any longer hold water? Brown *et al.* (in press) argue very strongly that it does not:

As we approach the end of the twentieth century, a gap has emerged in the analysis. The theorists do not say what happens when developing countries get trapped in the second stage, unable to achieve the economic and social gains that are counted upon to reduce births. Nor does the theory explain what happens when second-stage population growth rates of 3 per cent per year (which means a twenty-fold increase per century) continue indefinitely and begin to overwhelm local life-support systems.

The implications of this are clear if we consider the extent to which the world can now be divided into two groups: countries where population growth is slow or non-existent and where living conditions are improving, and countries where population growth is rapid and living conditions are often deteriorating. Once populations expand to the point where their demands begin to exceed the sustainable yield of forests, grasslands, croplands, or aquifers, they begin directly or indirectly to consume the resource base itself.

One of the best indicators that we are already up against some of the physical limits of Planet Earth is the work carried out by Vitousek and his colleagues, showing that the human economy uses—directly or indirectly—about 40 per cent of the net primary product of terrestrial photosynthesis today (Vitousek 1986). This implies that in a single doubling of the world's population (from 5.5 billion to 11 billion people by roughly the middle of the next century), we may well be using 80 per cent of terrestrial photosynthesis. Just where do we suppose we will eventually decide it is appropriate to draw the line, to allow at least some other species a modicum of breathing and living space on this planet?

To match the terrestrial limit there is a very clear atmospheric limit, in terms of the inability of the atmosphere to absorb an ever-increasing amount of the greenhouse gases we are emitting, particularly carbon dioxide (CO_2) and methane. With the exception of the United States and some of the Middle East oil producers, there is not a country on Earth that does not now subscribe to the international scientific consensus that global warming is a problem already with us and inexorably worsening.

But knowledge of these (and many other) limits have not yet persuaded world leaders that current economic and political ideas are wholly unsustainable—in the true meaning of that word, namely that they are clearly causing us to exceed natural carrying capacity. There is no mystery about this impasse: those same politicians cannot free themselves of their utter obsession with conventionally measured economic growth.

Economic growth is the measure of increased economic activity in any one year. And theoretically at least, that is all it is; but rather than acknowledging the somewhat humble limitations to the concept and pursuit of economic growth, almost all politicians since the Second World War have progressively built up its conceptual significance so that, instead of it being *one of* the means by which we achieve our ends, it has become the single most important end of human society.

The process is somewhat akin to obsessive obesity: instead of being a means of assuaging hunger, providing energy, ensuring convivial company or pleasant aesthetic experiences, the business of eating for such obsessives becomes the *only* activity which provides any kind of meaning or satisfaction or fulfilment. Eating itself becomes the goal of their existence. All industrial economies are now terminally confused between means and ends, between eating for living and eating for eating's sake.

Economic *growth* is simply not up to measuring economic *welfare*. It gives no indication of the sustainability of growth, and cannot measure the genuine efficiency of growth. Furthermore, it cannot indicate just *who* is benefiting from the growth, and cannot even discriminate between the genuine benefits of industrial production and the costs incurred in the process.

Sadly, we are trapped by our own dependency on economic growth, not as a *means* to an end, but as an *end* in itself—indeed almost as a new Godhead, before which we all dutifully genuflect. Every conceivable subterfuge and intellectual dishonesty is engaged in by politicians to avoid having to come to terms with this uncomfortable truth. Scientific evidence about the direct impact of exponential economic growth on our life-support systems (with more people inexorably demanding more from a strictly finite resource base and pollution absorption capacity) is dismissed as temporary or of secondary importance.

If sustainability is to be the *key* economic concept of the future, then the use of gross national product as the exclusive measure of economic success has obviously had its day. It is a bit like trying to assess one's enjoyment of a piece of music by measuring the number of notes in it!

Robert Goodland and Herman Daly (both of the World Bank) have succinctly mapped out the political dilemma that this obsession with conventionally measured economic growth has given rise to (Goodland *et al.*, in press).

Two realisms conflict. On the one hand, political realism rules out income redistribution and population stability as politically difficult, if not impossible. Therefore, the world economy has to expand 'by a factor of five or ten' in order to cure poverty. On the other hand, ecological realism accepts that the global economy has already exceeded the sustainable limits of the global ecosystem, and that a five- to ten-fold expansion of anything remotely resembling the present economy would simply speed us from today's long-run unsustainability to imminent collapse. We believe that in conflicts between biophysical realities and political realities, the latter must eventually give ground. The planet will transit to sustainability: the choice is between society planning for an orderly transition, or letting physical limits and environmental damage dictate the timing and course of the transition.

Indeed! But at the moment, the amount of serious planning for 'an orderly transition' is negligible. One can find no clearer example of this than the current UK energy policies—I use the word 'policies' advisedly, as the UK has no energy *strategy* inasmuch as this particular Government thinks that long-term strategic planning is ideologically wicked and an affront to its free market purity. With almost astounding impertinence, the Government confounds this wilful short-termism by referring to its ragbag of energy policies as being compatible with the overall goal of sustainable development. Rarely has the concept been more perniciously abused.

For Mr Wakeham and Mr Heseltine, sustainable development is an exceptionally useful platitude. It masks inaction, it masks a lack of vision and political courage, it masks total incoherence even within the Government's own ranks. Consider the case of domestic boilers.

The European Commission has proposed tough efficiency standards for all gas boilers as one of the easiest and most effective ways of achieving the EC's target of stabilizing CO_2 emissions at 1990 levels by the year 2000. But the proposals that they initially came up with were judged by Britain to be much too rigorous, inasmuch as only 40 per cent of British boilers would have actually met them.

Far from seeing this as an appropriate challenge to boiler-makers in the UK, the British Government argued and argued for a reduction in the actual standards. Ultimately, in order to get something agreed, the rest of Europe was bullied into agreeing to our lower standards. The direct consequence of this, put in quantitative terms, is that instead of a saving of 5 million tonnes of CO_2 per annum from the UK through more efficient boilers, all we shall now be achieving is a saving of 1.8 million tonnes of CO_2 per annum.

As a Department of Energy spokesperson said at the time, the Government was 'determined to improve energy efficiency, but not at a disproportionate cost and effect to the consumer'.

Or, worse yet, consider the case of energy efficiency. We are shortly due to see an EC-wide scheme to provide energy labelling for most domestic appliances. But everyone knows that such schemes only have a limited effectiveness, owing to a combination of ill-informed sales staff, complacent manufacturers, and muddled consumers. The obvious answer (accepted even by that free market purist former President Reagan) is for governments simply to impose minimal efficiency standards for all such appliances. Manufacturers who cannot meet those standards by a certain date simply will not be allowed to sell them.

From such fiascos, you would hardly know that the *official* Government policy is indeed to stabilize CO_2 emissions at 1990 levels by 2005—five years later than the rest of Europe. As it happens, rather than falling, CO_2 emissions in the UK have risen (by roughly 1 per cent per annum) over the last two years, at a time of profound economic recession when energy use would normally be expected to decline.

In the face of such hard evidence, the likelihood is, yet again, that the UK will be dragged reluctantly into some kind of 'sustainable transition' by the EC. The Commission's proposal for a carbon or energy tax is certainly the most radical yet to be put on the table, coming in at US$3 a barrel in 1993, and rising by US$1 a barrel for the next seven years to hit a figure of US$10 a barrel by the year 2000. The tax is calculated partly on the basis of the total energy content of whichever fuel is used, and partly on the specific carbon content. Clearly, the tax would dramatically increase coal, gas, and petrol prices by the end of the decade. But it is important to point out that the Commission's proposal ensures that this tax would be introduced as a substitute tax rather than an additional tax, ensuring 'fiscal neutrality' by allowing each member country to determine which of its existing taxes should be reduced proportionally.

The Commission estimates that this would lead to no more than a 0.1 per cent per annum decline in economic growth, but this has not stopped industrialists complaining vociferously about the threat it would pose to their competitiveness (*vis-à-vis* places like the United States, Japan, or the newly industrializing countries who are not introducing a similar tax), nor has it stopped the UK Government from declaring the whole thing to be 'premature'.

It is of course easy to be critical about this particular Government, but I am by no means convinced that any other government would be doing that much better. The Labour Party, for instance, has declared itself to be unequivocally opposed to any energy tax within Europe. The truth is that the incoherence and double-speak within both of the major parties arise from trying to graft the principles of sustainability onto an otherwise unchanged growthist ideology. And it simply cannot be done.

Again, the UK's energy policies exemplify this dilemma. I well recall the efforts made by Friends of the Earth during the course of the Electricity Privatization Bill to persuade the Government to amend the Draft Licence for the Regional Electricity Companies so that they would be permitted to sell *efficiency* as well as selling electricity. The idea of selling 'negawatts' rather than megawatts is already well established in the United States, where many utilities have seen energy saving as a far more effective way of meeting demand than building new generating capacity.

It should be perfectly possible for the Regional Electricity Companies to do the same, for instance by distributing energy efficient light bulbs *for free* to as many of their consumers as possible, and then passing the costs of that apparent 'freebie' on to consumers via the regular electricity bills. But such straight-forward market mechanisms were considered inappropriate by this Government—for the simple reason that the financial fortunes of the Regional Electricity Companies are driven entirely by their ability to sell more and more electricity. For it is on that (as well as on increased productivity) that the fortunes of their shareholders depend.

The Government's continuing obsession with nuclear power provides another sorry example. Study after study has revealed that nuclear power was one of the least cost-effective ways of achieving a reduction in CO_2 emissions. Every pound spent on one option is a pound not spent on another. As Mrs Thatcher stressed in her address to the Royal Society in September 1988, 'no nation has unlimited funds, and it will have even less if it wastes them.' Hear hear! Given that investment in energy efficiency is at least five times more cost effective as a way of reducing CO_2 emissions, it is truly remarkable that anyone should still be looking to nuclear power as a solution.

Jimmy Carter once referred to the energy crisis as the 'moral equivalent of war'. Any world leaders worth their salt (particularly if they happen to be scientists) would already be planning for a very different kind of future. Looking ahead to the year 2025, when the oil begins to run out and sea levels begin to rise, they would surely conclude that the greater the proportion of our energy needs that can be provided for by renewables, the better off we would be. They would then presumably decide, at a stroke, to do for renewables what the post-War politicians, for very different reasons, did for nuclear power: direct the necessary human and financial resources into research and development programmes with the best chance of coming up with the answers, not tomorrow or even the day after, but between now and that not so distant crunch point in the next century.

At the very worst, large sums of money would be wasted on technologies that simply could not be made to work or proved to be too expensive (like a re-run of our prolonged experiment with nuclear power). But at least no one would be killed, and very little pollution would be caused. At the very best, internationally coordinated development programmes over the next thirty years would crack the outstanding technological problems, and provide humankind with unlimited, non-polluting sources of energy from then on.

The Government's current target for renewables is for them to provide up to 20 per cent of the UK's energy by 2025. The short-term target is 1000 megawatts by the year 2000—roughly 2 per cent of our current electricity needs. Under the first two tranches of the Non-Fossil Fuel Obligation (which supports renewable energy producers only as a fortunate by-product of supporting nuclear power), the Government has already given the go ahead to more than 550 megawatts, which is not a bad start at all.

But in reality, it is still small beer. Here are just some of the things that we could and should be doing to ensure that renewable power really becomes the mainstay of our whole energy supply system within the first two decades of the next century.

1. Set up a new Renewable Energy Authority and attract to it the best brains in the business from all over the world.

2. Totally reverse the balance of the existing research and development budgets between nuclear power (around £165 million) and renewables (around £12 million).

3. Initiate a massive wind energy programme, with a view to generating up to 20 per cent of our electricity from onshore wind generators by 2025. Simultaneously establish the first major UK offshore wind farm, in the North Sea, with a reasonable expectation of being able to generate more and more of our electricity in this way in the next century.

4. Commission a tidal barrage (with financial support from the Government) across the Mersey, to come on stream before the year 2000, and complete a detailed feasibility study on all other potential sites in the UK for similar barrages: Morecambe Bay, the Solway Firth, the Wash, the Dee, the Humber, and the Severn.

5. Provide incentives to make maximum use of potential 'biofuels': agricultural wastes can be treated in specially designed digesters to generate gas that could be used for heat or electricity; straw and forestry wastes can be burned in specially designed furnaces, and 'surplus agricultural land' could be planted with fast-growing trees specifically for energy use.

6. Instantly restore funding to our wave power programme. In the longer term, offshore deep water wave energy is an immensely exciting prospect, and we should aim to have several large-scale prototypes in operation within the course of the next decade.

7. Aim to achieve a target of 5 million passive solar installations (involving roughly a quarter of current UK housing stock) and at least 5 million solar water heaters—bearing in mind that Japan already has more than 3.5 million of them. Set up

a special team to conduct further research into the development and application of solar cells with a view to reducing their cost from $5 per watt to $1 per watt, at which point they will be in direct competition with electricity from the grid.

As an afterthought, I hope these wise and far-seeing politicians will also set up a special Renewable Energy Export Agency, not only to recoup some of the considerable costs entailed in this kind of development programme, through massive export sales, but also to ensure that benefits of these renewable technologies are shared as widely as possible with developing countries. Solar cells are already a highly economical electricity source for Third World villages, and a far higher proportion of their needs will certainly be met from this source in the next century. Indeed, it is clear that a large proportion of the energy needs in the Third World can only be met from this and other renewable sources.

I will not for a moment pretend that this course of action would be either easy or cost-free. Indeed, I have made it clear on countless occasions that without some profound shift in our *values* (as well as in our choice of technologies and our political priorities), it is highly unlikely that we will ever make peace with the planet. The 'have your cake and eat it' brigade offers us the seductive possibility that all we have to do is to learn to consume a bit more sensitively, and all will be well with the world. Well it won't. Green consumerism may marginally slow down the erosion of our life support systems, but it cannot possibly stop the process altogether.

And that of course is the political crunch. To achieve sustainability, we in the rich North have to reduce the level of material throughput in our economies. We have to achieve 'development without growth and throughput beyond environmental carrying capacity'. When something grows, it gets quantitatively bigger; when it develops, it gets qualitatively better. This is an essential principle on which any 'new deal' between rich North and poor South totally depends.

Again, Robert Goodland and Herman Daly put their finger

on the political reality beyond this (Goodland *et al.*, in press):

It is neither ethical nor helpful to the environment to expect poor countries to cut or arrest their development, which tends to be highly associated with throughput growth. Therefore the rich countries, which are after all responsible for most of today's environmental damage, and whose material well-being can sustain halting, or even reversing, throughput growth, must take the lead in this respect. Poverty reduction will require considerable growth, as well as development, in developing countries. But ecological constraints are real, and more growth for the poor must be balanced by negative throughput growth for the rich.

Such an agenda would be exceptionally difficult to implement. Markets, for example, will have to learn to function without expansion, without wars, without wastes and without advertising that encourages waste. Economic policy will have to suppress certain activities in order to allow others to expand, so that the sum total remains within the bio-physical budget constraint of a non-growing throughput. This adds up to a formidable political agenda. That is why exceptional political wisdom and leadership are so urgently required.

Reduced material throughput (which, in political terms, should be understood to mean a reduced material standard of living) in the North, coupled with urgent programmes of debt relief and income redistribution, is the quid pro quo of seeking reductions in population growth rates in the Third World.

I would be the first to agree that this is not the easiest political message to take to an electorate brainwashed over the last forty years by cornucopian illusions and daft promises of permanently increasing material well-being from *all* political parties except the Greens. But it can and *must* be done.

Only in this kind of context does the concept of sustainability become genuinely useful. The dishonest and platitudinous definitions from political leaders are dangerous, for they draw a veil over the real state of the Earth today, and postpone the time when we stop playing at being a little bit 'cleaner and greener', and undertake with resolution and imagination the profound trans-formation of our industrial society that is now so urgently required.

REFERENCES

Brown, L., Chandler, W. V., Flavin, C., Pollock, C., Starke, L., and Wolf, E. C. (1991). *State of the world: a Worldwatch Institute report on progress toward a sustainable society*. Norton, London.

Goldsmith, E., Allen, R., Allaby, M., Daroll, J., and Lawrence, S. (1972). *Blueprint for survival*. Penguin, London.

Goodland, R., Daly, H., and el Seraphy, S. *Environmentally sustainable economic development: building on Bruntland*. (In press.)

IUCN/WWF/UNEP (1991). *Caring for the Earth: a strategy for sustainable living*. Earthscan, London.

Meadows, D., Meadows, D., Randers, J., and Behrens, W. (1972). *The limits to growth*. Earth Island Press, London.

Meadows, D., Meadows, D., and Randers, J. (1992). *Beyond the limits*. Earthscan, London.

Vitousek, P. (1986). Human appropriation of the products of photosynthesis. *Bioscience*, June 1986.

World Commission on Environment and Development (1987). *Our common future* ('The Bruntland report'). Oxford University Press.

3

Clean power from fossil fuels

Peter Chester

After taking a First in physics at Queen Mary College, London, and completing his research for a Doctorate, Dr Peter Chester spent seven years in North America, first as a Post-doctoral Fellow with the National Research Council in Ottawa and then at the Westinghouse Research Laboratory in Pittsburgh. In 1960 he returned to the United Kingdom to set up the Solid State Physics Section in the Central Electricity Research Laboratories (CERL); six years later he was appointed Research Manager (Sciences) at the Electricity Council's Research Centre. Dr Chester moved back in 1970 to the Central Electricity Generating Board (CEGB), where he served successively as Controller of Scientific Services (NW Region), Director of the Central Electricity Research Laboratories, Director of the Technology Planning and Research Division, and, finally, Director for the Environment. When the CEGB was dismantled as a result of the privatization of the electricity generating industry, Peter Chester was appointed Executive Director for Technology and Environment on the Board of National Power plc. In 1984 he gave the Institute of Electrical Engineers Faraday Lectures and in 1985 was awarded the Robens Coal Science Medal. He has written extensively in the fields of solid state and low temperature physics and on energy and the environment. In 1992 he was elected a Fellow of the Royal Academy of Engineering and to an Honorary Fellowship of Queen Mary and Westfield College. He is currently Chairman of National Wind Power Limited.

INTRODUCTION

Other chapters in this volume are concerned with nuclear power and renewable energy sources. This one will focus on the conversion of the energy stored in fossil fuels into useful power and the prospects for doing so with greater environmental acceptability.

For those not technically inclined, it will first attempt to bring out as simply as possible the fundamental principles that limit the efficiency of the energy conversion process, then outline the engineering advances that have been made within those limits, the avenues for further progress, and some of the unsolved problems.

HEAT ENGINES AND THEIR LIMITATIONS

Fossil fuels are composed mainly of carbon and hydrogen. Their latent chemical energy is released as heat during combustion—a process in which these elements react with oxygen to produce carbon dioxide (CO_2) and water vapour. This thermal energy can be converted to mechanical energy, which in turn can perform useful work or generate electrical power, by using some form of 'heat engine'.

One of the earliest heat engines was the reciprocating steam engine, followed by the steam turbine, the internal combustion engine, and more recently by the gas turbine, which revolutionized mass travel and is about to do the same for power generation.

Clearly, the higher the efficiency of a heat engine, the lower its fuel consumption for a given duty and the lower the emissions from combustion. Fortunately for the environment, the efficiency of heat engines has been improving steadily, but will it ever be possible to achieve 100 per cent efficiency?

The answer to that question is found in the Laws of Thermodynamics—three fundamental truths of Nature which cannot be evaded. In popular terms they can be expressed as follows:

First Law: 'You can't win. You can only break even.'

Second Law: 'You can only break even at Absolute Zero.'

Third Law: 'You can't get to Absolute Zero.'

The First Law means that it is impossible under any circumstances to get more energy out of the engine than is in the fuel in the first place. A statement of the obvious perhaps, but consistently ignored by aspiring inventors of perpetual motion machines. The second law needs a little more explanation.

Consider a car engine. When the spark-plug fires at the top of the piston stroke, a hot, high-pressure ball of gas is formed. The hotter it is, the greater its energy. This hot gas pushes the piston down and makes the engine turn. In the process of doing this work, the pressure of the gas falls and so does its temperature. The bigger the expansion, the more work can be extracted from the gas and the greater the fall in temperature. But there is still some energy left in the hot exhaust gas and that energy goes to waste.

Thermodynamics tells us that the energy in an 'ideal' gas is proportional to its 'Absolute Temperature' as measured from the 'Absolute Zero' of temperature, i.e. minus 273 °C. That relationship is represented by the straight line in Fig. 3.1.

It was the French scientist Sadi Carnot who first realized in 1824 that even a perfect heat engine using an ideal gas has to reject some of the input heat energy. The fraction rejected is equal to the ratio of the exhaust temperature to the initial temperature, both measured from Absolute Zero. It will be obvious from Fig. 3.1 that only if the final temperature were actually Absolute Zero could all the energy in the gas be converted to work. Since the Third Law of Thermodynamics precludes reaching Absolute Zero, it is impossible to convert all the heat into useful work, even with a perfect working fluid and a device that is perfect in every mechanical respect.

The Third Law is not a real limitation for energy conversion since in practice the exhaust temperature cannot be lower than that of the cooling system that carries away the reject heat.

Ever since Carnot, therefore, engineers have striven for the highest achievable initial temperatures for their heat engines and for exhaust temperatures as close as possible to ambient. No real gas is ideal and no real heat engine is perfect. So in practice even the Carnot efficiency is unattainable. The extent to which it can be approached depends on the skill of engineers in devising working

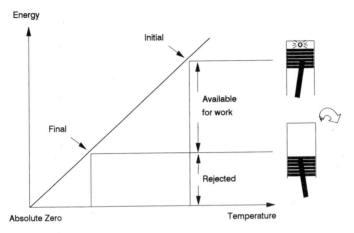

Fig. 3.1 The temperature and energy relationship in a perfect heat engine.

cycles for real fluids and designing systems that will survive high temperatures and pressures with minimum incidental energy loss.

The gas in the cylinder of a car engine immediately after ignition starts with a temperature of about 2000 °C and exhausts at about 400 °C. Its Carnot efficiency is therefore about 70 per cent. In practice, limitations of weight, materials, and cost reduce this figure to 40 per cent or less for a diesel engine and 30 per cent for a spark ignition engine. The waste heat appears in the exhaust gases and in the air warmed by the radiator. These figures relate to optimum steady test speeds. Much lower efficiencies would be returned under urban driving conditions.

Early steam engines had efficiencies of only 5 per cent or so. Modern coal-fired power stations are eight times as efficient.

EMISSIONS FROM COMBUSTION

Coal contains mineral materials which survive the combustion process as fine particles of ash. In a modern station some 99.8 per cent of the ash is captured in electrostatic precipitators before the flue gas is released to the atmosphere. Oil also contains ash but in much smaller quantities, generally not requiring controls.

The main gaseous emission is carbon dioxide—a major contributor to the 'greenhouse effect'. The amount emitted per unit of heat released depends on the proportion of carbon to hydrogen in the particular fuel. Natural gas produces only 56 per cent of the carbon dioxide released by coal for a given heat output.

A common impurity in coal and oil is sulphur which is oxidized to sulphur dioxide in the combustion process. In British coals sulphur generally ranges from 1 per cent to 3 per cent by weight. Other coals, for example some from America and Australia, range between 0.5 per cent and 0.8 per cent. Refinery residual oils that are burned in power stations range up to 3.5 per cent. Natural gas, as supplied, is virtually sulphur free.

When any fuel is burned in air, which is four-fifths nitrogen, some of that nitrogen is oxidized in the flame to form oxides of nitrogen, NO_x for short. The higher the temperature of the flame, the more nitrogen is oxidized and the higher the NO_x levels. In addition, fossil fuels contain up to 1 per cent nitrogen in their composition and some of this is oxidized during combustion.

THE STEAM CYCLE

The simple steam cycle is shown schematically in Fig. 3.2. Powdered coal is burned in a stream of air to give a flame temperature of about 1500 °C. About 90 per cent of the heat of combustion is captured in a boiler which produces steam under pressure and raises its temperature. It is not possible to get steam as hot as the flame because there is no practical material that could stand the temperature and pressure. The practical limit is about 570 °C.

The superheated, high-pressure steam, impinging on the blades of a steam turbine, causes it to rotate and drive an electrical generator. The steam gives up some of its energy at each stage of

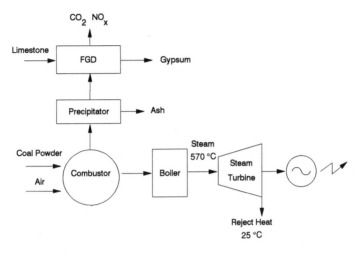

Fig. 3.2 The simple steam cycle for electricity generation with flue gas desulphurization (FGD).

the turbine as it passes through, finishing up as condensed water at close to ambient temperature.

The Carnot efficiency of this steam cycle would be 64 per cent. It follows from the Second Law that such a station must reject at least 36 per cent of the input energy as low temperature heat. That is no fault of the designers (unless you include the Almighty in that description). It is to their credit that currently, practical cycles with real materials can achieve 42 per cent efficiency, i.e. two-thirds of the theoretical maximum.

'End-of-pipe' controls for the steam cycle

As coal and oil have given way to electricity for motive power in factories and to gas and electricity for heating, urban levels of dust and sulphur dioxide have fallen steadily, typically by half in the past decade. In contrast, urban nitrogen oxide levels have risen over the same period by about the same percentage, owing to the burgeoning growth of road transport and our congested streets.

By the 1970s, the discovery of long-range transport of air pollutants taught us that we must be as much concerned for the distant environment—indeed for the global environment—as for the local environment. Attention was first focused on reducing emissions of sulphur dioxide from industrial boilers and power stations. There appeared to be no practical way of removing sulphur from coal before combustion and efforts therefore concentrated on washing it from the flue gas after combustion. The process is called flue gas desulphurization, or FGD for short.

There are several forms of FGD. The 'limestone–gypsum' system being installed at Drax Power Station in North Yorkshire is the most common type in Europe. After passing through the precipitators, the flue gas is blown through a spray or cascade of water made alkaline with powdered limestone—a natural form of calcium carbonate. The reaction with sulphur dioxide produces calcium sulphate, recovered as a whitish powder better known as gypsum and widely used in wallboard manufacture. There is also a gaseous product from the limestone, carbon dioxide, which adds to the emissions from the station.

At Drax, the scrubbing towers are each 48 m high and the duct work feeding them some 9 m in diameter. It has taken considerable ingenuity to shoehorn the enormous ducts and towers into the existing station layout. The work will take seven years and will cost some £700 million.

The reaction removes 90 per cent or more of the sulphur dioxide but there is an environmental price to pay. The yearly limestone requirement is half a million tonnes and the annual production of gypsum 800 000 tonnes. In the process 200 000 tonnes of additional carbon dioxide are released. The energy requirements of the process reduce the efficiency of the station from about 39 per cent to about 38 per cent.

In contrast to sulphur dioxide, the release of nitrogen oxides is as much a function of the combustion conditions as of the fuel. The first line of attack on nitrogen oxides therefore begins with the flame. If the burner is designed to restrict the mixing of fuel and air so as to starve the initial stages of combustion of oxygen, there are two benefits. First the flame temperature is reduced, thus reducing the oxidation of atmospheric nitrogen. Second, the nitrogen atoms in the fuel have more chance to combine with each other instead of with an oxygen atom.

'Low NO_x burners' designed on this principle typically reduce NO_x levels by half. Similar measures are used in gas turbines by careful design of the combustion chambers or by introducing steam during combustion.

A further reduction of NO_x can be achieved after combustion by introducing ammonia into the flue gas which reacts with the NO_x to form water and nitrogen. Commonly a catalyst is used to promote this reaction and this is carried on the surface of a large bed of ceramic honeycomb or pellets through which the gas is blown. The process is known as selective catalytic reduction or SCR and it is capable of reducing NO_x by 80 per cent. The capital cost of a retrofit to a 2000 megawatt (MW) coal-fired station would be upwards of £100 million with operating costs in addition.

The catalytic converter on a car exhaust works on a similar principle, using the carbon monoxide generally present in the exhaust, rather than ammonia, to react with the NO_x.

Processes such as FGD and SCR, designed originally as retrofits to existing plants, have become known as 'bolt-on' or 'end-of-pipe' controls. They are generally sub-optimal in performance, costly, often difficult to retrofit, and reduce the efficiency of the energy conversion process to a certain extent.

THE COMBINED CYCLE GAS TURBINE

Further improvements in the efficiency of the simple steam cycle are difficult to achieve because of the temperature and pressure limitations of alloys for the boiler tubes and steam turbine casings. But there is another avenue—use of the combustion gases themselves as a working fluid in a gas turbine.

The principle of the gas turbine is first to compress air in a rotating compressor, mix it with natural gas under pressure in a combustion chamber, and burn it. The resulting flame is both hot and at considerable pressure. The hot gases emerging from the combustion chamber drive the turbine blades directly. Although the temperature is higher than in a steam turbine, the pressure is lower and within the capability of suitable alloys.

Clearly, the greater the compression that can be achieved in the compressor stage and the higher the gas temperature that can be tolerated by the turbine stage, the greater the potential efficiency of energy conversion. These have been precisely the goals of the aero-engine manufacturers over the past four decades. The power industry is now able to benefit from them.

In the latest designs, the gas turbine itself operates with an inlet temperature of 1250 °C. Not all of the energy in the hot gas can be extracted in the gas turbine and it typically leaves the turbine at a temperature of 500 °C—hot enough to run a conventional steam cycle.

This has led to the combination of gas turbines and steam cycles, each with its own generator, in a 'combined cycle gas turbine' (CCGT), as shown schematically in Fig. 3.3. Such systems currently achieve overall conversion efficiencies of 54 per cent—two-thirds of the theoretical maximum—and are being supplied on a production line basis by several major manu-facturers. Because of the purity of the natural gas supply, there is virtually no emission of sulphur dioxide, SO_2.

It is therefore not surprising that CCGTs are the energy con-version technology of choice wherever natural gas is available. Figure 3.4 shows the relative impact of a CCGT compared with a coal-fired station of the same capacity with FGD.

Fuel consumption is cut by 25 per cent and CO_2, NO_x, and waste heat by about 60 per cent. Solid waste and the requirement for limestone are eliminated and sulphur dioxide virtually so. Construction is quicker, less disturbing to the neighbourhood, and at half the cost of a new coal-fired station. As an investment in environmental improvement, the CCGT is seen as the current 'best buy' for the green consumer and the cost-conscious con-sumer alike.

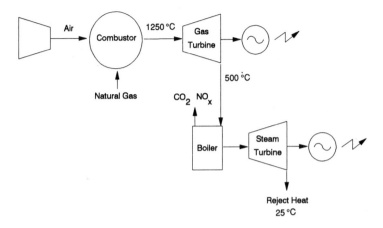

Fig. 3.3 The combined cycle with gas turbine and steam turbine (CCGT).

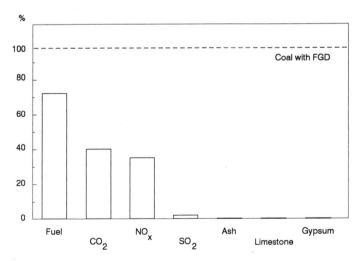

Fig. 3.4 The relative environmental impacts of combined cycle gas turbine generation and coal-fired power generation with FGD.

The first independent CCGT in the UK came on stream at Roosecote in 1991 and there will be further CCGTs commissioned throughout the decade, possibly amounting to 10 000 MW in all.

There is, of course, no prospect that natural gas can supply all the world's electricity needs. Even with the projected programme of CCGTs in the UK they will supply no more than a quarter of UK electricity demand. Gas is fairly promoted by British Gas as a 'Bridge to the Future' but that future must inevitably include a major role for coal.

ADVANCED COAL CONVERSION TECHNOLOGY

What are the prospects of more efficient and cleaner conversion technology for a coal-based future? A number of new concepts are under development or at the demonstration stage. The common feature is to convert the coal to gas, either by combustion or gasification, and to use a gas turbine in some form of combined cycle. This section will outline the main technical thrusts, their ultimate advantages, and the environmental challenges they still pose.

The pressurized fluidized bed combined cycle, PFBC

The pressurized fluidized bed cycle (Pillai 1989) is shown schematically in Fig. 3.5. The fuel takes the form of crushed coal. Air is compressed and introduced at the base of a bed of burning coal at such a speed that the bed is fluidized. This promotes good combustion and good heat transfer to the boiler tubes immersed in the bed. Crushed limestone added to the bed reacts with sulphur dioxide to form gypsum, as in the FGD process, but this now remains behind with the ash. The hot combustion gases, still under pressure, need to be cleaned of dust in high temperature filters before they pass into a gas turbine which drives a generator.

The exhaust from the gas turbine raises further steam in a heat

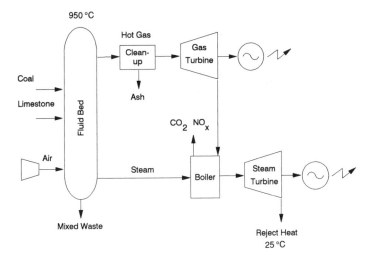

Fig. 3.5 The pressurized fluidized bed combined cycle (PFBC).

recovery boiler which is converted to electricity in the conventional way with a steam turbine and second generator.

The maximum temperature of the fluid bed is limited to about 950 °C—anything hotter would cause the ash to melt and the bed to seize up. For this reason, the efficiency of this cycle is limited to about 41 per cent.

Sulphur removal can be as high as 90 per cent and NO_x levels are relatively low. However, the quantity of solid discharge from the fluidized bed is greater than that from a conventional plant and is more difficult to dispose of—consisting of ash mixed with gypsum and excess quicklime arising from the limestone. This is a leachable material, requiring impermeable linings at disposal sites.

Three demonstration plants are under construction but because of the restriction on bed temperature, the potential of this cycle is limited.

Integrated gasification combined cycle, IGCC

The temperature limitation of the PFBC can be avoided by converting the coal to a fuel gas in a gasifier. Gasification is a process that has been used for many years in the petrochemical industry to provide feedstock for chemical synthesis. Over this time three main types of gasifier have evolved (Peters 1989). The most common type is known as the entrained gasifier which itself has been developed in several different commercial forms.

The essentials of the process are shown in simplified form in Fig. 3.6. Powdered coal (or oil) is partially burned in a restricted stream of oxygen and steam under pressure to produce a fuel gas consisting mainly of carbon monoxide and hydrogen and containing up to 80 per cent of the initial energy in the fuel. The gas can then be cooled and cleaned using well proven commercial processes before being fed into a combined cycle gas turbine. By integrating all the waste heat flows between the gasifier, the

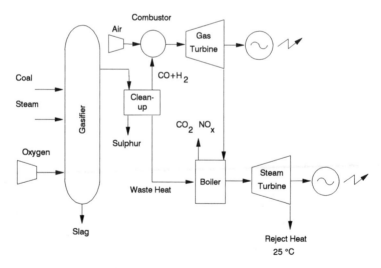

Fig. 3.6 The integrated gasification combined cycle (IGCC).

turbines, and the boiler, an overall efficiency of close to 45 per cent can be obtained with current technology. Hence the name for this cycle—integrated gasification combined cycle.

A demonstration IGCC, to be commissioned at Buggenum, The Netherlands, in 1993, is expected to have an efficiency of 42.5 per cent, while a later demonstration at Puertollano in Spain in 1996 will possibly reach 45.5 per cent. Under present emission standards, neither is expected to be fully commercial in competition with coal using current end-of-pipe controls. Nevertheless, the environmental advantages of IGCC are very substantial in comparison with other coal options.

A particular advantage of the IGCC is the fact that its solid waste appears as a glassy slag, highly resistant to leaching and acceptable as a construction material. Impurities in the fuel gas can be reduced to almost any desired level, the sulphur being extracted as elemental sulphur for which there is a market. Production of NO_x in the gas turbine combustion chamber can be reduced to low levels by suitable design.

Most IGCCs are based on oxygen and therefore need an oxygen separation plant. This raises the capital cost and makes it harder for IGCCs to compete with CCGTs. For this reason British Coal and others have sought ways of circumventing the temperature limitation of the cheaper PFBC in other ways. These are commonly referred to as 'topping' cycles.

'Topping' cycles

The 'topping' cycle concept takes a number of forms, none of which has yet reached demonstration stage. A common principle underlying all of them is to gasify the coal partially in air rather than in oxygen. One concept is illustrated in Fig. 3.7. The fuel gas consists mainly of carbon monoxide, hydrogen, and nitrogen. After a high temperature clean-up process, still to be developed, the gas is burned in a high-temperature gas turbine. The char residue from the partial gasification passes into a pressurized fluid bed combustor which raises further steam for the steam

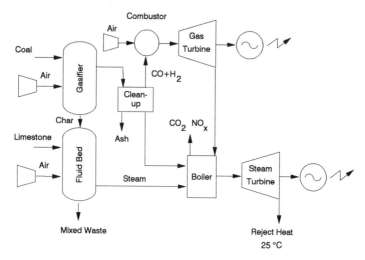

Fig. 3.7 An example of a 'topping' cycle.

cycle. In this way it is anticipated that efficiencies of 45 per cent may be achieved.

Environmentally the cycle is not as attractive as the IGCC. As with the PFBC the solid waste is a leachable mixture of ash and gypsum together with unreacted lime.

ENVIRONMENTAL COMPARISONS

In Table 3.1 the various technologies are compared in terms of their emissions per unit of electricity generated. From it some important conclusions can be drawn:

● Even the best 'clean' coal technology on the drawing board will not reduce CO_2 emissions per unit of electricity produced by more than 15 per cent compared with a modern conventional station with FGD, even if all the technical hurdles can be overcome.

Table 3.1. Comparative discharges (in grams), per unit of electricity generated (kilo-watt hour), from different types of power station

	Sulphur dioxide (g/kWh)	Nitrogen oxides (g/kWh)	Carbon dioxide (g/kWh)	Solids (g/kWh)
Conventional coal				
with FGD	1.2	2.3	830	75
future with FGD	1.0	2.3	780	
future with FGD + SCR	1.0	0.3	780	
Fluid bed cycles				
simple	1.1	1.2	830	90
circulating combined	1.2	1.2	820	
pressurized combined	1.0	1.1	780	
future 'topping'	0.9	0.7	710	
Coal gasifier cycles				
IGCC	0.1–0.5	0.8	760	55
future IGCC	0.1–0.5	0.8	730	
CCGT	–	0.8	380	None

- IGCCs have the greatest potential for NO_x and SO_x reduction.
- IGCCs produce the least solid waste and it is insoluble in ground water.
- Most significant of all, only CCGTs are capable of substantial reductions in CO_2 emissions.

COMBINING HEAT AND POWER

In all these cycles the exhaust temperature is deliberately kept as close as possible to ambient to maximize the efficiency of electricity generation. The inevitable reject heat is then too cool to be of much use. It is, however, possible to design cycles that exhaust at a higher temperature—high enough to provide useful process heat or steam or space heating, as depicted in Fig. 3.8. This is known as 'combined heat and power', CHP, or cogeneration.

Inevitably this produces less electricity per unit of fuel consumed but the heat from the CHP plant displaces fuel that would

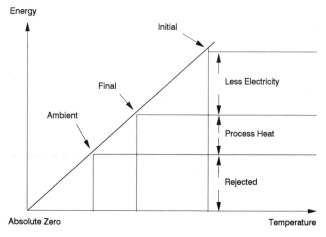

Fig. 3.8 The temperature and energy relationship in an ideal combined heat and power (CHP) system.

otherwise be used to provide it. When account is taken of this displaced fuel, the overall efficiency of a cogeneration system can be as high as 85 per cent.

The most common heat engines for larger systems are the gas turbine and, below 1 MW, the diesel engine. In either case the exhaust heat is used to raise process steam or to heat water for space heating. The electricity output is used by the owner of the plant, any excess being exported to the local electricity company who may also top up the supply at times of peak demand.

Trading electricity in this way is a great deal easier than trading heat, especially where there are multiple end-users such as in a district heating scheme. This requires extensive pipework systems, large capital investments, and the conjunction of a large number of different interests.

In contrast, cogeneration in single-site situations such as factories, hospitals, hotels, and office blocks is practical and capable of effecting useful reductions in CO_2 emissions (Evans 1990). The rate of penetration of such systems is presently limited by the need for new capital investment in the plant and the rate of return expected on it. The barriers to progress are not necessarily technical.

Self-evidently, cogeneration has little application in the conversion of fossil energy to motive power for transport and we must look elsewhere for a breakthrough in this regard (Okken 1991).

DIRECT ENERGY CONVERSION

So far, this chapter has concentrated on the conversion of energy with heat engines and the limitations of this principle. Could one not extract the chemical energy from fuel directly, without the need to convert it into heat, and thus circumvent the intrinsic and practical inefficiencies of heat engines?

There is nothing new in this concept—indeed it is as old as life itself. Our own bodies convert the chemical energy in carbohydrates and fat into motive power by oxidation processes that depend on electrochemistry rather than combustion. The carbon dioxide and part of the water vapour we breathe out are the end products of these oxidation reactions.

It was not until 1839 that the chemists caught up with Nature when William Grove invented the fuel cell. It had been known for some time that when a metal is in contact with an electrolyte such as an acid, charges are transferred at the interface with the result that the metal acquires an electrical potential, or voltage, with respect to the acid. Each metal has its own characteristic potential. So, if two different metals are immersed in the acid, there will be a potential difference between them. Connect the two with a wire and a current can flow. This is the familiar principle of the battery. In the process one or both of the electrodes is consumed by the reaction.

Grove's innovation was to replace the metals with porous electrodes fed with hydrogen and oxygen respectively. Since these gases have different potentials in relation to the acid, a voltage is set up between the electrodes which can drive a current and do useful work. In the process, charged hydrogen atoms cross the cell and are oxidized at the oxygen electrode to form water. To promote the reactions, the electrodes must be coated with a suitable catalyst—in this case platinum or other precious metal. Fuel cells of this kind have reached a high degree of technical sophistication and an overall efficiency of about 48 per cent.

Fuel cell technology has reached an advanced stage (Appleby 1992). Environmentally they are extremely clean because of the high purity required of the input gases. The only waste product of the cell itself is water. However, the hydrogen is produced from methane, methanol, or other hydrocarbon fuel by a process which inevitably produces carbon dioxide and in some cases NO_x as well.

Fuel cells are not cost competitive with central power generation at this stage of development but they may become so for embedded power generation in heavily loaded urban distribution systems. If effective use can be made of the waste heat in a combined heat and power application, the overall efficiency could rise to 80 per cent. A number of demonstration systems are in progress in Japan and the USA.

Two other types of fuel cell have greater potential for co-generation. The electrolyte in the first of these is a molten mixture of carbonates and, therefore, has to operate at 600 °C. The fuel is again hydrogen produced from methane by a reformer, which in this case can be thermally integrated with the fuel cell itself. The overall efficiency, gas to electricity, is about 60 per cent.

These cells are not yet commercial for power generation. Development work is aimed at reducing costs, and increasing efficiency and life. The high operating temperature lends itself to the effective use of waste heat so that ultimately cogeneration efficiencies of 80 per cent can be anticipated.

A more distant prospect is a cell based on a ceramic electrolyte operating at 1000 °C. These can decompose natural gas without a reformer and make use of carbon monoxide as well as hydrogen as fuel. Efficiencies around 55 per cent can be expected.

Because of their potential efficiency and low emissions of NO_x, fuel cells are of interest for transport applications (Ascoli 1991). The favoured type uses a special polymer as electrolyte and hydrogen, from reformed methanol, as fuel. It operates at about 90 °C with an efficiency, methanol to electricity, of about 30 per cent. One major difficulty, yet to be solved, is its intolerance to carbon monoxide.

LIMITATION OF CARBON DIOXIDE EMISSIONS

In our review so far we have seen that it is, or will be, technically possible to remove at least 90 per cent of all the emissions of concern, *except carbon dioxide*. The best prospects here, at the present time, are the greater use of CHP, switching from coal to natural gas, and radical improvements in efficiency. The CCGT combines the last two. In terms of immediate impact on carbon dioxide emissions, every 1000 MW of base-load CCGT displacing coal-fired generation currently results in a 3 per cent reduction in CO_2 emissions from the British generating network or, alternatively, would allow it to cope with a 3 per cent growth in demand for no increase in CO_2.

It will not be so easy for those countries that depend heavily on coal for their economic growth. Annual emissions of carbon dioxide from the former Soviet Bloc, and those from China and India, have been growing by about 5 per cent per annum. Greater energy efficiency in these countries is clearly needed. But, if current predictions of global warming are sustained, the world will eventually need to use coal without releasing carbon dioxide into the atmosphere. That is possible in principle if the CO_2 is separated out and stored.

Coal and oil without carbon dioxide?

Studies of the possibility of removing CO_2 from power stations are well advanced (Hendriks *et al.* 1989), and the first international conference on the subject was held in Amsterdam in 1992. There are a number of separation processes based on technologies already in existence in the chemical industry which could be integrated into the design of future coal-fired stations. The best starting point would be gasification in oxygen to produce carbon monoxide and hydrogen as in the IGCC, followed by steam reforming to produce more hydrogen and carbon dioxide (Hendriks *et al.* 1991). These authors estimate that the cost of electricity from an IGCC would increase by about 25 per cent for 88 per cent CO_2 removal, compared with an increase of 68 per cent from a conventional coal-fired plant.

There are at least three different processes that might be used to separate the carbon dioxide from the hydrogen before it is burned—absorption of CO_2 in a solvent or chemical reagent, compression or refrigeration to condense the CO_2, or separation with a diffusion membrane. All these techniques are currently subject to paper studies to evaluate relative merits and costs.

A number of studies have been made of the practicality and

cost of disposing of the separated CO_2. Van der Harst and van Nieuwland (1989) and Hendriks *et al.* (1991) conclude that depleted gas wells in The Netherlands could store large quantities, possibly the equivalent of forty times that country's annual CO_2 production, at a cost less than that of separation.

Van Engelenburg and Blok (1991) have examined the possibility of storing CO_2 in aquifers—deep beds of sand capped with a stratum of rock and permeated with saline water. Such geological formations underlie many parts of The Netherlands. The concept is to displace the saline water with carbon dioxide under pressure. Their study suggests a storage capacity equivalent to 200 years of current Dutch CO_2 production, at a cost possibly lower than that of storage in depleted gas wells. A great deal more work will be required to prove the storage capacity and its ultimate integrity against leakage or escape.

Another possibility is to dissolve carbon dioxide in the lower layers of the ocean, below about 500 metres. The water at these depths is colder and denser than the surface layers and the currents in it take centuries to emerge at the surface. The storage capacity would be enormous and could provide an invaluable alleviation for many decades. A variant on this is to pump liquid CO_2 to the deep ocean below 3000 metres, or to drop frozen CO_2 to this depth, where it would form a stable pool of liquid, denser than the ocean itself. The turnover time at this depth could well be thousands of years. Much more work is needed to validate and cost these concepts and to assess their environmental acceptability (de Baar and Stoll 1989; Herzog *et al.* 1991).

A less radical and probably cheaper option could be to use only part of the carbon in coal in the first place. This could be done by partial gasification—as in the topping cycle—but returning the carbon-rich char to the mine for indefinite storage. This concept could well reopen studies of underground gasification—in this case partial gasification—leaving much of the carbon in the coal measure.

Hydrogen as a fuel?

If the large-scale storage of CO_2 were to become commonplace, hydrogen would become a major fuel. The concept of a hydrogen-based energy future, some of it produced by nuclear electricity, is not new (Lecocq and Furukawa 1991). Hydrogen gas can be transmitted readily through pipelines under pressure and could well form the basis of an industrial fuel network with only water as its combustion product. Liquid hydrogen in vast quantities already powers space rockets. For surface transport too, liquid hydrogen would be the most suitable form. A hydrogen-powered car has already been demonstrated in Germany.

CONCLUSIONS

The acid gases SO_2 and NO_x

For power generation adequate bolt-on technology is available now, at a price, for coal- and oil-fired stations. As an environmental investment for the future, arguably, it would be better to put the money into systems based on coal gasification because of their greater cleanliness and efficiency. However, combined cycle gas turbines using natural gas offer even better efficiency and environmental performance already and at a lower cost.

The 'greenhouse' gas, carbon dioxide

CCGTs available today produce far less CO_2 per unit of electricity than the best 'clean' coal technology on the drawing board now or likely in the future.

If greenhouse fears are sustained, natural gas will truly be 'a Bridge to the Future' and the prospects for coal itself could well depend on the extraction and storage of carbon dioxide or carbon itself. Ultimately, if carbon dioxide extraction is necessary on a significant scale, hydrogen may emerge as the cleanest fuel, distributed in pipelines or in liquid form.

Cogeneration

CO_2 emissions can be reduced to some extent by improving the efficiency of energy conversion, whatever the fuel. However, by far the greatest gains will come from greater use of cogeneration and CHP. This is more readily achieved in single-user applications close to the heat load. For the future, fuel cells look well suited to this kind of application fed with a pipeline gas—either natural or from a gasifier or hydrogen.

The internal combustion engine

Adequate end-of-pipe technology exists for NO_x reduction to European standards. There is little prospect for dramatic reductions in CO_2 from efficiency improvements but a switch to compressed natural gas as fuel would help, as could electric vehicles and less travelling. Eventually we may see hydrogen-powered vehicles.

A caution

Technology thrives on challenges such as these and no doubt much of this could come to pass. However, the greenhouse issue is a global one. The capital investments implied are of mind boggling order and would cripple the economies of the developing world. What is best for the West may be quite wrong for them. Many of them have other resources amenable to intermediate technology—biomass and energy forestry included.

Harmonization of technical approach could well be counterproductive. In protecting the global commons we must not let the lure of high technology run away with our common sense.

REFERENCES

Appleby, A. J. (1992). Fuel cell technology and innovation. *Journal of Power Sources*, **37**, 223–39.

Ascoli, A. (1991). Fuel-cell powered electric vehicles: overview and perspective. *International Journal of Global Energy Issues*, **3** (4), 217–20.

de Baar, H. J. W. and Stoll, M. H. C. (1989). Storage of carbon dioxide in the oceans. In *Climate and energy* (ed. P. A. Okken, R. J. Swart, and S. Zwerver), pp. 143–77. Kluwer, Dordrecht.

Evans, R. D. (1990). Environmental and economic implications of small scale CHP. *Energy and Environment Paper No. 3*. Energy Technology Support Unit, Harwell, UK.

Hendriks, C. A., Blok, K., and Turkenburg, W. C. (1989). The recovery of carbon dioxide from power plants. In *Climate and energy* (ed. P. A. Okken, R. J. Swart, and S. Zwerver), pp. 125–42. Kluwer, Dordrecht.

Hendriks, C. A., Blok, K., and Turkenburg, W. C. (1991). Technology and cost of recovering and storing carbon dioxide from an integrated-gasifier, combined-cycle plant. *Energy*, **16**, 1277–93.

Herzog, H., Golomb, D., and Zemba, S. (1991). Feasibility, modelling and economics of sequestering power plant CO_2 emissions in the deep ocean. *Environmental Progress*, **10**, 64–74.

Lecocq, A. and Furukawa, K. (1991). Hydrogen production for the next century. *International Journal of Global Energy Issues*, **3** (4), 204–9.

Okken, P. A. (1991). A case for alternative transport fuels. *Energy Policy*, **May 1991**, 400–5.

Peters, W. (1989). Coal gasification technologies for combined cycle power generation. In *Electricity* (ed. T. B. Johansson, B. Bodlund, and R. H. Williams), pp. 665–95. Lund University Press.

Pillai, K. K. (1989). Pressurised fluidised bed combustion. In *Electricity* (ed. T. B. Johansson, B. Bodlund, and R. H. Williams), pp. 555–93. Lund University Press.

van der Harst, A. C. and van Nieuwland, A. J. F. M. (1989). Disposal of carbon dioxide in depleted natural gas reservoirs. In *Climate and energy* (ed. P. A. Okken, R. J. Swart, and S. Zwerver), pp. 178–88. Kluwer, Dordrecht.

van Engelenburg, B. and Blok, K. (1991). Prospects for the disposal of carbon dioxide in aquifers. Report from the Department of Science, Technology and Society, University of Utrecht, The Netherlands, February 1991.

4

The gas industry:
survivor and innovator

Denis Rooke

Sir Denis Rooke, CBE, FRS, FEng. has devoted his working life to the British gas industry. A determined champion and influential spokesman for it, he has also gained a reputation as an expert in gas matters world-wide. He joined the South Eastern Gas Board in 1949 as a Mechanical Engineer and was subsequently seconded to the North Thames Gas Board to work, both in the UK and in the USA, in the field which was to trigger the industry's subsequent technological revolution, liquefied natural gas. In 1959 he was a member of the technical team on board the Methane Pioneer, *which brought the first cargo of liquefied natural gas to Britain. Appointed Development Engineer to the Gas Council in 1960, he played a seminal role in the transition to an integrated natural gas industry, including the development of a national high-pressure gas grid to exploit the new North Sea resources and the development of the industry's own explora-tion and production arm. He became the full time Member for Production and Supply of the Gas Council in 1966, Deputy Chairman in 1972, and Chairman of what then became the British Gas Corporation in 1976. Holding the Chairmanship until 1989, Sir Denis guided the industry through its successful privatization and became the first Chairman of British Gas plc.*

Sir Denis was elected to the Fellowship of Engineering (now The Royal Academy of Engineering) in 1977 and served as President 1986–91. Elected a Fellow of the Royal Society in 1978 and awarded the Society's Rumford Medal in 1986, he has been Chancellor of Loughborough University of Technology since 1989.

INTRODUCTION

This chapter is concerned with the important, positive impact on environmental issues, past, present, and future, made by the gas industry.

Although environmental concerns today are global, observed global effects are the sum of many individual local events. If practical ways of influencing the overall situation are to be found, an understanding of these local events must also be achieved. Therefore it is principally interactions with the environment by the UK gas industry, and the users of its products, that are addressed in this chapter.

ENVIRONMENTAL ISSUES

There has been a marked change in the perception of environmental issues in recent times. The realization of their significance has been a gradual process, although it has accelerated in recent years. No longer are ill-informed pressure groups solely responsible for lobbying. Now lobbyists are usually well informed, and the general public is becoming increasingly involved. The public seems to consider protection of the environment as important, and is asking industry to provide much more information. What is not yet known is to what degree the customer is prepared to pay for more stringent environmental control. As is well known, costs rise very steeply when tighter control is applied to remove, or eliminate, the last traces of pollutant.

The economics of pollution control must necessarily be considered carefully by business. The Centre for Environmental Science and Technology has estimated recently (CEST 1991) that UK expenditure associated with environmental issues over the next decade could reach £140 billion, £50 billion of it on the greenhouse effect alone. To justify such a level of expenditure considerable economic benefit would need to be demonstrated.

Alongside the increased involvement of the public there has

been a long-term broadening of concern over local pollution issues to include environmental issues with national, international, and global dimensions. There is concern, not simply for what is happening in the country of residence, but also for locations much further afield, such as destruction of rain forests, protection of whales or seals, or the potential exploitation of resources in sensitive areas such as Antarctica.

Nowadays, many companies in the gas sector are in business on an international scale, and are thus subject to a wide range of environmental requirements. British Gas, which has only been privatized since 1986, now operates on a global scale in over forty countries, and is under the same scrutiny as other multinationals. The use of poor environmental practice in less developed countries, or remote locations, has increasingly become less acceptable. Companies involved in enterprises in such places are constantly reviewing their policies and practice against new requirements, although potential changes still need to be assessed within the framework of economic viability.

THE DEVELOPING GAS INDUSTRY

Coal gasification

Gas has a history which goes back beyond the eighteenth century and its development has been well documented by Hugh Barty-King (1984) and Trevor Williams (1981). Pioneering work was carried out by John Clayton who described the distillation of coal, storage of the gaseous product in a bladder, and its lighting properties.

Later a Scottish engineer, William Murdoch, working for Bolton and Watt in Cornwall, developed the basic idea further and designed permanent gas lighting systems. It was 200 years ago this year that he demonstrated the use of coal gas to light a room in Cross Street, Redruth. He was awarded the Royal Society Rumford Gold Medal for his work; it was appropriately inscribed 'Ex Fumare Dare Lucem' (to give light out of smoke).

However, no attempt was made to patent his process. It was a German, Frederick Windsor, who won the initial battle for exploitation in this country.

Although the presence of flammable gases in nature was well known, the production of gas in this country for the following 150 years was based on heating coal to a high temperature, out of contact with air, in horizontal, vertical, or inclined retorts. In the nineteenth century this was essentially a batch process. It was physically arduous and labour intensive; even when mechanical aids had been designed it was wasteful in energy because of the water or steam quench of incandescent coke, and the need to keep retorts at red heat.

In early practice the apparatus was somewhat unsophisticated, and working conditions were unpleasant. In the raw gas, methane and hydrogen accounted for approximately 85 per cent and carbon monoxide for 5 per cent, but small quantities of other toxic or objectionable gases, such as hydrogen sulphide, ammonia, and hydrogen cyanide, were present. Careful treatment of the gas before distribution was therefore necessary, and tar and ammonia scrubbed from the raw gas became important by-products of the industry. In addition, the sulphur removed became the basis of part of the sulphuric acid industry.

As time passed improvements in the carbonization process brought greater mechanization of many operations, as well as substantial increases in thermal efficiency and, overall, a welcome improvement in working conditions and environment. The desirability of continuous processing was fully appreciated and the vertical retort came to dominate the scene, with the coal fed continuously at the top and coke removed at the bottom. One of the constraints was that the coal feed had to meet tight specifications if the coke bed was not to collapse, and the handling of huge tonnages of solids, both coal and coke products, still led to a considerable dust nuisance.

In order to increase the efficiency of the process, a possible modification in the operation of a vertical retort was to add steam at the base to produce a carbon monoxide and hydrogen mixture (water gas), although that could lead to an increase in aqueous liquor. Alternatively, some of the product coke was gasified in automatic cyclic plant, more flexible than carbonizing plant, to make water gas. Coke oven gas was also purchased from the steel industry.

At first the development of town gas was purely for the supply of gas for lighting, both of premises and public places. It is worth remarking that public gas lighting was responsible for a considerable social revolution in making the streets safe for citizens after dark, thus adding greatly to the opportunities for public concourse. The early lighting appliances relied upon the luminosity of the flame, a function of its temperature and the presence of solid carbon particles, until the invention of the incandescent mantle by Welsbach in 1885. This in fact made the luminosity of the gas quite irrelevant; the important thing was to make the flame as hot as possible, but it was only in 1920 that the unit of gas quality changed from candle power to calorific value.

Although Frederick Windsor had foreseen the benefits of gas for cooking and heating, it was only during the latter half of the nineteenth century that the practice became at all widespread. More and more manufacturers marketed small domestic appliances, cookers, water heaters, normally called geysers at that time, and space heaters—both convection gas heaters and radiant gas fires.

Industrial use of gas commenced with small-scale operations such as welding and brazing, but later large quantities were absorbed in heavy industries such as steel, brick, and glass making. Figure 4.1 shows the distribution of gas between market sectors in 1990–91 when industrial gas accounted for 29 per cent of sales.

The post-war years and nationalization

The years of the Second World War were difficult for the gas industry. As young men were drafted away into the services, the labour intensity of the carbonization process led to increasing labour difficulties. So much so that women were drafted in to

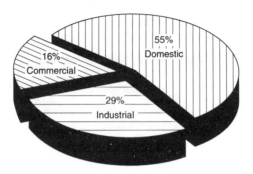

Total = 19.6 million therms

Fig. 4.1   Gas sales in 1991.

carry out maintenance work. Shortage of materials added to the problems of effective maintenance.

In addition, there were too many small gas-making units. At nationalization in 1949, 1044 gas works were in production. No less than 605 of these had an annual output less than that required to supply 800 average domestic customers of today.

Twelve Area Boards were established under a central Gas Council and charged with the task of rationalizing the industry. But to add to the difficulties, the industry found itself the captive of the coal industry. The National Coal Board controlled all the supplies of coking coal and could selectively price it. Furthermore, the principal by-product of carbonization was coke which had to be disposed of as a domestic fuel in competition with other, and cheaper, grades of coal. It was a classic case of economic squeeze, the situation being made even more difficult because of the siting of oil refineries in the UK, and Europe, after the Second World War. These provided additional competition in the non-transport energy market from refinery by-products. Finally, electricity could be generated from coals costing less than the specialized types required for gas making. As a result, sales of gas began to decline. Over ten years the numbers of works in production were sharply reduced so that by 31 March 1960, only 169 of the smallest category remained out of a total of 428 in production. Despite these changes the economic position of the industry worsened appreciably.

Gas from oil

By 1955 the Gas Council had long recognized the inherent weakness of the position described above and saw the economic advantages of total gasification, preferably at pressure, and better still, of the use of an oil feedstock instead of coal. It had established its Midlands Research Station at Solihull, under Dr F. J. Dent, FRS, who possessed unrivalled knowledge and understanding of the control of the complex reactions possible between hydrogen, carbon, and oxygen. By mixing and matching several processes, fuel gas with a range of calorific values and of acceptable density and flame speed, critical for correct combustion in existing appliances, could be obtained.

Numerous gasification processes, at first using heavy oil, were being developed in different parts of the world and one or two examples of many of these new cyclic processes were built in the UK. But the Onia-Gegi process, a low-pressure, cyclic, gasification plant, was the dominant process until the availability of cheap petroleum naphtha, a volatile hydrocarbon distillate, opened up wider possibilities.

It was a stroke of luck that the pressure for improvement in urban air quality, arising from the Clean Air Act of 1956 and associated Smoke Orders, occurred at a time when the gas industry was fighting for a new image and new products. As new processes for gas manufacture were being developed, the chemical industry was also active in seeking a route to cheaper bulk hydrogen. ICI developed a high-pressure catalytic steam-reforming plant to convert light petroleum distillate into a synthesis gas as the first step in the production of methanol or ammonia for fertilizers.

The gas industry was quick to appreciate the potential of the new process, notwithstanding that the calorific value of the product gas was too low to allow it to be distributed directly as town gas. It needed to be enriched and that is where the novel high-pressure gasification processes developed by Dr Dent proved ideal. By adding these enrichment processes to an ICI reforming plant, town gas could be manufactured at high pressure, from low-cost feedstock, in large low-cost plants with minimal manning requirements. The first such plant was commissioned in 1963, and by the end of 1967, 200 individual reforming plants were on stream.

One very ingenious enrichment plant was the Recycle Hydrogenator in which hydrogen, in the lean gas product from an ICI reformer, was reacted with petroleum distillate at high pressure and temperature (750 °C). This process, entirely non-catalytic, was widely used.

A second, more interesting process was the Catalytic Rich Gas Process which involved a unique and highly active catalyst. This adiabatic process was conducted at a pressure of 30 bar and at around 500 °C, and converted light petroleum distillate and steam directly to a methane-rich gas. This could be used to enrich synthesis gas, or was itself used as a reforming feedstock giving easier reaction conditions and a wholly self-sufficient plant. This process led, in due course, to the design of plants to make a fully compatible substitute natural gas (SNG).

The natural gas era

It was around this time that another significant development occurred. There were large, known reserves of natural gas in the world, much of it being flared during the production of oil. Indeed, countries such as the USA already had a natural gas industry based on national reserves.

Following pioneering work on the transport of liquefied natural gas (LNG) to the UK in a specially modified ship, the Government authorized the importation of LNG from Algeria in purpose-designed tankers. (One volume of LNG provides 600 volumes of gas on evaporation.) The long-term availability of this source of natural gas led to construction of the first high-pressure pipeline between the LNG terminal at Canvey Island and Leeds with supply spurs to eight Area Boards, or Regions as they have been called for many years now. The natural gas was used to enrich lean reformer gas. Sales of gas began to increase dramatically.

Following the discovery of natural gas in The Netherlands, and subsequent gas discoveries in the southern North Sea, decisions about the future of the gas industry had to be taken. Should there be a big capital investment in further reforming plant, with the associated running costs? Should the calorific value be raised beyond approximately 19 MJ m^{-3} (megajoules per cubic metre) or 500 Btu ft^{-3} (British thermal units per cubic foot) to another value, for example approximately 28.5 MJ m^{-3} (750 Btu ft^{-3}). Increased demand suggested that increased storage would also be required. A change to natural gas had attractions. It would double the storage capacity purely by virtue of its higher calorific value, and the natural high pressure of the supply would enhance transmission and distribution capacity.

The main difficulty in the direct utilization of natural gas was that burners needed to be altered to supply the same quantity of heat in a given time, with a flame of the same size and shape as before. However, this was a technology which had been well researched by the industry and the decision to move to the direct supply of natural gas was taken. The then Chairman of the Gas Council, Sir Henry Jones, announced in June 1966 that the whole country would be converted in the decade 1967–77. It was an immense operation which involved visits to over 13 million premises. Thirty-five million appliances had to be converted and 650 million parts were required. The cost of £600 million was funded entirely by the gas industry which, in addition, also had to fund an extra £400 million arising from the early retirement of redundant production plant.

This revolution in the industry also required the laying of a greatly expanded length of high-pressure pipeline operating at up to 75 bar. This now provides a national energy transmission system which has continued to function effectively through hurricanes and blizzards when other forms of energy supply have broken down.

For diurnal peaks gas holders still provide significant storage. Some people obviously find these landmarks attractive and in fact the holders at St Pancras are the subject of a preservation order. However, others cannot wait to see gas holders disappear from the landscape. As holders come to the end of their useful life, they will not be replaced. Nowadays, other facilities such as high-pressure storage are used.

To meet the large seasonal peaks in demand for gas a number of approaches are used. Increased flow can be called from contracted supplies, and from the Morecambe Bay Field which is owned by British Gas and can thus be regulated as required. In addition, gas can be taken from the Rough Field, an offshore storage reservoir into which gas is pumped during the summer and produced in large quantities during the peak periods in winter. Large salt cavities approximately 2000 m below ground, which have been leached out by British Gas, are also used for gas storage to help meet peak winter demands. Each cavity can hold approximately 44 million m^3 of usable gas. Finally, gas is taken continuously from the transmission grid at certain locations, liquefied and stored in above ground cryogenic storage tanks to be re-evaporated to meet sudden swift increases in demand.

It was the combination of all these elements that led to the successful growth of the gas industry in the UK which at the same time yielded positive contributions to an improved environment. Somewhat similar developments have taken place in many Western European countries.

THE BENEFITS OF NATURAL GAS

Air quality

We all would prefer to live in a safe and improving environment, and for that reason the use of fossil fuels has for many years been critically assessed in terms of their potential for environmental damage. The main concern initially was focused on health effects, particularly those associated with smoke or particulates.

Those who are old enough will remember what heavy urban smoke was like and how the use of clean forms of energy led to a very significant improvement in the quality of urban air. Figure 4.2 also shows that an improvement in the urban levels of sulphur dioxide has been achieved in recent years as gas usage increased whilst end-use energy remained almost constant. These improvements in air quality occurred not only in Britain, but in cities in Germany and other European countries. The installation

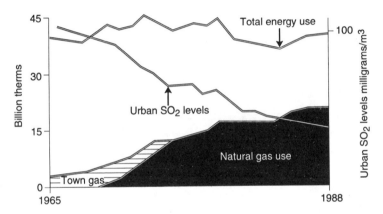

Fig. 4.2 Reduction of urban sulphur dioxide (SO_2) levels compared with increasing use of natural gas. (From Department of the Environment Digest of UK Energy Statistics, HMSO, 1990.)

of gas systems to improve air quality is still occurring in major cities such as Ankara in Turkey, and there are many towns in Eastern Europe and further east which have serious pollution which is not yet being tackled effectively. Indeed, from a global viewpoint, it would be more cost effective to invest in pollution control in those locations as opposed to seeking marginal improvements in more developed neighbouring countries.

Subsequently, the emphasis has moved from particulate control to consideration of other pollutants such as nitrogen oxides (NO_x), ozone, and volatile organic compounds, including hydrocarbons. This control of pollution was considered a national matter until, some twenty years ago, Scandinavian countries began to draw attention to the increased acidity of their lakes. The acid rain debate had begun.

International experts designed large research programmes, including that sponsored through the Royal Society, to study atmospheric chemistry, the phenomena of wet and dry deposition, and the effects of deposition on water quality, vegetation, soil chemistry, and materials. The results of those studies have shown that practically nothing is as simple and straightforward as had originally been supposed.

However, as a result of the continuing debate and political negotiations, many countries, including the UK, have agreed to reduce emissions into the air. Fossil-fuel use has been the primary target for such emission control.

In this respect natural gas has significant advantages in comparison with other fossil fuels. Figure 4.3 shows that natural gas use in different market sectors yields very low levels of sulphur dioxide

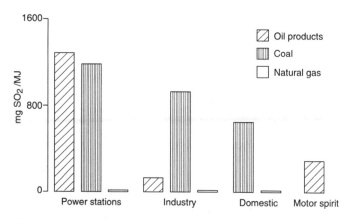

Fig. 4.3 Sulphur dioxide emissions from different market sectors.

per unit of energy used. As already mentioned, it does not produce particulates. Natural gas also has advantages in relation to the emission of nitrogen oxides during combustion of fossil fuels because, unlike coal or oil, it does not produce NO_x from nitrogen combined within the combustible fuel molecules.

Global warming

Just as opinion was becoming adjusted to the acid rain debate and its international dimension, attention was drawn to global environmental problems. Ozone depletion is one effect generating concern, even in Europe, but the debate on global warming, in which the absorption of long-wavelength radiation from the Earth is postulated to give rise to an enhanced greenhouse effect, has perhaps been the issue of more direct relevance to the gas industry.

Large national and international co-ordinated research programmes, on a whole range of factors and effects, are in place and are still being augmented. It is, of course, a very complicated subject and there are still many uncertainties. A full understanding may take years to acquire but, in the meantime, a precautionary approach to remedial measures is justified on general efficiency grounds if the cost is not disproportionate.

The main trace gases, apart from carbon dioxide, that contribute to the greenhouse effect are methane, chlorofluorocarbons, nitrous oxide, and ozone. Their activity in this respect and their lifetimes in the atmosphere vary enormously, and it is the integration of these two factors which is relevant. Methane has a very short lifetime compared with carbon dioxide.

The magnitude of carbon dioxide emissions and their rate of increase have focused additional attention on the use of fossil fuels. In addition to work of the Intergovernmental Panel on Climate Change (Houghton *et al.* 1990) there have been many other, sometimes more local, studies. A report by the Watt Committee (Thurlow 1990) found that, in the domestic, commercial, and industrial sectors of the UK economy, substantial savings in carbon dioxide emissions, of approximately 20 per cent, could be gained by the implementation of measures that were both technically proven and cost effective. Similar conclusions on the possibilities for improved energy efficiency were reached in the UK Department of Energy Paper No. 58 (1989).

When it comes to global warming, natural gas has an advantage over other fossil fuels in that the amount of carbon dioxide released, per unit of energy used, is lower owing to the higher hydrogen to carbon ratio. So, as a fuel, taken with other factors, it has benefits of low environmental impact overall. This benefit might find use in other ways. For example, road traffic is recognized as being responsible for a large and increasing fraction of national pollution inventories of carbon monoxide, non-methane hydrocarbons, and NO_x (Department of the Environment 1990). Figure 4.4 shows how

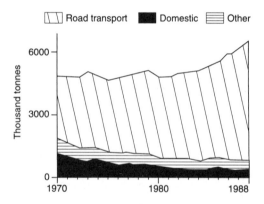

Fig. 4.4 Carbon monoxide emissions in the UK. (From Department of the Environment Digest of UK Energy Statistics, HMSO, 1990.)

significantly carbon monoxide emissions have grown. In addition, road transport contributed up to 45 per cent of the national NO_x emission in 1991. Whilst legislation to reduce that contribution is in hand there is also a marked increase in smoke from poorly adjusted heavy diesel vehicles. One possible remedial measure for pollution from traffic is the use of natural gas vehicles (NGVs).

Such vehicles can have much lower emissions and could be especially beneficial for urban fleets. Comparative results shown in Fig. 4.5, from vehicles that have been converted to take natural gas, indicate that there are already immediate potential benefits from NGVs, and these are likely to be greater when the engines are purpose designed.

It is interesting to recall that it is by no means the first time that gas has been used to fuel cars; it was practical using both coal gas and producer gas during the Second World War.

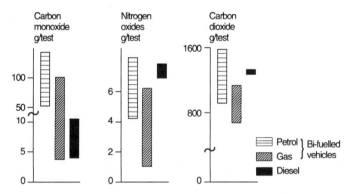

Fig. 4.5 Natural gas vehicle (NGV) emission data—preliminary UK results. (Reproduced from IGE communication 1974, Institution of Gas Engineers, London, 1991.)

MINIMIZING ENVIRONMENTAL IMPACT

Air quality

It is not sufficient that the product should rely solely on inherent advantages. It is essential to ensure that the industry carries out its operations with the minimum impact on the environment. In particular, an understanding of the science of pollution, and how pollution affects air quality, must be maintained. For that reason

British Gas established in 1987 what is believed to be the first urban acid rain and air quality monitoring unit in the country. Of course it is also necessary to take measurements of air quality in the field. For this type of monitoring it is essential to have mobile units which carry meteorological and pollutant monitors, including units to measure methane, sulphur dioxide, and NO_x. British Gas has such a unit which is easily transferred to a ship for work offshore, or used on a more conventional chassis for work on the road system.

Ecology and pipeline reinstatement

The advent of the natural gas era required a new approach to the laying of pipelines because most of the construction was not in towns, as in former days, but in rural settings. Transmission pipelines up to 1050 mm in diameter had to be laid in a way that would have minimal permanent impact on the countryside.

In the beginning it was often sufficient to reinstate pipeline routes to the satisfaction of the landowner; that, in effect, meant reinstatement of land for grazing or arable use. Nowadays there are pressures to seek ways of reinstating the original habitats. For example, in some countries there has been anxiety expressed over the loss of wetlands, and ecologists have been studying methods that might help achieve the objective of habitat reinstatement.

Within British Gas particular attention has been paid to ensuring effective reinstatement following pipeline installation in sensitive habitats such as salt marshes. Much preconstruction work is carried out in surveying the area and detailing precautions to minimize damage to the habitat. Once engineering work is completed significant care is taken to returf, avoid erosion, and prevent invasion by alien species.

Considerable effort has also been devoted to satisfactory restoration of heathland. The topsoil is typically very thin with 95 per cent of the seed bank and nutrients in the top 50–100 mm, and unless a careful programme is devised there can be a complete change in the character of the reinstated land. Techniques have been developed to cause the minimum disturbance to soil and vegetation. A similar approach has also been applied in boggy areas which are particularly sensitive to disturbance.

Visual impact

The development of the natural gas supply system also involved consideration of the visual impact of new installations. The task of placing shore terminals, compressor stations, and storage facilities at locations to meet operational requirements, and yet satisfy environmental constraints of protected areas, required great skill in environmental planning. Specialized architects were invariably consulted to ensure that plant designs would be of the highest visual standard.

Most of the new installations, such as compressor stations at strategic locations along the pipelines, are in rural surroundings and have to meet stringent noise limits as well as visual constraints. Visual impact was not dealt with by merely 'painting it green'. Architects helped to evolve colour combinations that were often strong but nevertheless in harmony with the rural surroundings, or ensured that the units would merge into the horizon when viewed from afar. In other cases, for example LNG storage, it is not possible to disguise the presence of large storage tanks. However, great care was taken to ensure that colour variations were employed, for example using horizontal bands of blue on tanks on one site, which would merge with the skyline.

Ensuring water quality, on and offshore

Discharges to water have always been a contentious environmental issue, and attention to effluent treatment was one of the topics which received early attention by the gas industry. More recently British Gas has been closely involved in hydrocarbon exploration and production in the UK. In the 1970s the Gas Council was successful in developing the Dorset oilfields in an area criss-crossed with areas of ecological importance, and demonstrated that it could be done successfully. Protection of water quality was a prime concern.

What has been a new departure in recent years is the anxiety not only in Europe, but in the USA and elsewhere, regarding the potential impact of offshore exploration and production on the marine environment. British Gas owns and operates two gas fields in waters near the English coast—the Morecambe Field in the Irish Sea and the Rough Storage Field in the North Sea, and in addition it is associated with further ongoing exploration activities in European offshore waters and elsewhere.

One of the subjects of most concern is contamination from drilling muds. In drilling its exploration and production wells, British Gas endeavours to use water-based muds, as far as geological conditions and safety requirements permit. The Morecambe Bay development is close to the shore and the area has major inshore shell fisheries. This was a major consideration in the decision not to use oil-based muds. However, the Rough Field had to be developed using oil-based muds owing to the nature of the rock formation, and their use ensured greater hole stability and improved safety.

Twenty per cent of the oil input to the North Sea is believed to arise from exploration and production activities which are now subject to monitoring. However, before any such requirement was in place, British Gas established a research programme to study the potential impact of its activities in the Irish Sea, and subsequently expanded that programme to encompass the Rough Field also. The programme set out to collect sediments and benthic fauna, those small animals which live on the seabed, on a fixed pattern of sampling locations up to 6 km away from the drilling rigs. Obviously it is important to record the background conditions before exploration and production activities commence.

Figure 4.6 shows enhanced levels of barium around a Morecambe platform. They arise from drilling mud associated with drill cuttings discharged to the sea. The levels of heavy metals, Fig. 4.7, are low compared with the existing background concentrations, which probably came originally from the Mersey. As expected, there is little, if any, increase in hydrocarbon levels. It is interesting to note that a Ministry of Agriculture, Food, and

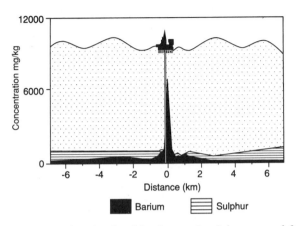

Fig. 4.6 Concentration levels of barium and sulphur around More-cambe Bay platforms.

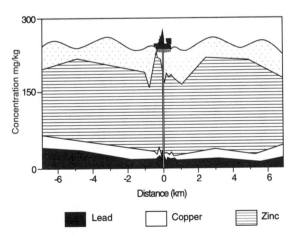

Fig. 4.7 Concentration of heavy metals around Morecambe Bay plat-forms.

Fisheries pre-drilling survey showed elevated hydrocarbon contents of 400 milligrams per kilogram, a situation which underlines the importance of carrying out baseline surveys.

In the Rough Field there is a much lower background of everything, including metals, because the sediment is much coarser than that in Morecambe Bay. Hydrocarbons from the oil-based muds have shown up in the sediments, but the year on year trend does not show accumulations significant enough to pose a problem. The coarse nature of the sediment has permitted rapid degradation to baseline conditions since drilling activity began in 1988.

The changes from the 1988 maximum levels to the 1989 levels, at several sampling points, can be seen in Fig. 4.8. Note that the

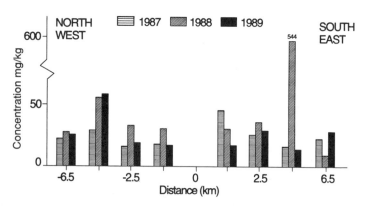

Fig. 4.8 Hydrocarbon concentration in sediment around the Rough Field platforms.

maximum concentrations do not occur at the platform but some distance from it. In a later survey concentrations had fallen even lower. The studies have shown that these operations are having only minimal environmental effects and, therefore, the operational practices followed offshore can be considered satisfactory.

It is also interesting to point out that in some situations operations need to be protected from the environment! The animals which give rise to marine fouling on platforms can cause significant problems of structural stability and solutions are constantly being sought. One approach is to study the life cycle of the animals so that their reproduction can be minimized or less harmful species can be encouraged to grow preferentially.

Efficient use of gas

We have seen that natural gas has many inherent advantages as a fuel in terms of its low emissions of NO_x and carbon dioxide, and negligible particulate and sulphur dioxide emissions. Nevertheless, an increase in the efficiency of energy use is also a desirable objective and can reduce emissions of pollutants still further. Increased efficiency in energy use is being sought in many countries, for example in relation to white goods, lighting, and cars, etc. Increased efficiency in use has been a specific objective of the UK gas industry for many years now.

To help achieve this objective British Gas established a school of fuel management for representatives from industry, and has encouraged the drive for fuel efficiency and savings through an annual Gas Energy Management award scheme in which companies compete each year. Annual savings by the entrants in 1991, in the sixteenth year of the awards, amounted to enough gas to provide the annual supply to approximately 45 000 average domestic customers.

In addition to these initiatives, the gas industry itself undertakes research into methods of achieving improved efficiency in the use of gas. This has involved, for example, the development of recuperative and regenerative burners, gas burning at the point of use to replace steam systems, condensing boilers, combined cycle gas turbines, and combined heat and power. The development of condensing boilers, in which heat is transferred from the flue gas through heat exchangers to the incoming water, has led to efficiencies of 85–90 per cent. The results of improved efficiency are reduced emissions of carbon dioxide, and of other pollutants, in many applications. At the same time it has been important to ensure, under the spirit of the best practical environmental option

(BPEO), that pollution is not transferred elsewhere, for example to water.

Power generation using natural gas offers significant advantages. The most efficient technique is the use of a combined cycle gas turbine in which there are two stages. Efficiencies approaching 50 per cent should be achievable. The combination of low carbon dioxide per unit energy, and high efficiency, means that the carbon dioxide emission from a modern combined cycle gas turbine system is halved compared with that from a modern coal power plant. In appropriate circumstances systems can be coupled to give even higher efficiencies, by using waste heat for drying, or to provide process heat.

Industry often has to balance pollution control against energy efficiency. This is illustrated in Table 4.1 which shows a comparison of a high-temperature application of three heating systems using natural gas (NG), with equivalent oil and electric systems. Going from the cold air burner using gas to the recuperative burner, the figures show that there is a 50 per cent increase in efficiency, from 20 to 30 per cent. However, to control the emission of NO_x requires a 3 per cent reduction in efficiency to a final figure of 27 per cent. Nevertheless, the gas recuperative burner is still more efficient than the other systems, and in addition the mass of nitrogen oxides and carbon dioxide per hour is lowest with the recuperative gas burner.

ENVIRONMENTAL MANAGEMENT IN THE GAS INDUSTRY

In recent years there has been an increased emphasis on improving environmental management within industrial organizations. In the UK the Confederation of British Industry has been taking a very positive line and encouraging companies to recognize that good environmental management can lead to increased profits too.

The gas industry in this country has already taken a positive

Table 4.1. Efficiency and NO_x emission for three heating systems

	Cold air burner	Recuperative burner (no NO_x control)	Recuperative burner (with NO_x control)	Cold air burner	Electric furnace
Fuel supply	NG	NG	NG	Gas oil	Coal-fired power station
Efficiency (%)	20	30	27	21	20
NO_x ppm (ref 3% O_2)	150	400	200	175	380
NO_x emitted (mg/h)	85	148	81	100	283
CO_2 emitted (kg/h)	54	36	32	72	98

stance on such issues. In June 1990 it published its Environmental Policy Statement. This Statement emphasizes the commitment to take full account of the environmental, health, and safety
implications of its operations, and to protect the natural environment. In particular, British Gas aims to comply with the spirit, as
well as the letter, of environmental health and safety legislation
and approved codes of practice, co-operating fully with relevant
statutory and non-statutory bodies to assess the likely environmental effects of planned projects and operations, and to maintain throughout its operations standards of environmental
protection reflecting best industry practice in comparable
situations, improving on such standards where reasonably
practical and economic. It also aims to foster among staff, suppli-
ers, customers, shareholders, and communities local to British
Gas operations, an understanding of environmental issues, and to
report publicly on the Company's environmental performance.

Wider debate on major environmental issues has been encouraged through sponsorship of a series of some fourteen short
papers by recognized experts on subjects such as renewable
energy (Elliot 1990), the urban environment (Button 1990), and
environmental education (McCleish 1990), and a more detailed
study of corporate responsibility for the environment (Pearce
1991).

British Gas recently published its first annual report (1991) on
environmental performance. It explains the principal issues
facing the company, reports on actions taken, and outlines plans
for the future. Actions already taken include the initiation of an
environmental audit, work to reduce uncertainty in natural gas
leakage figures, and a survey programme to check for off-site
pollution.

THE FUTURE ROLE FOR GAS

As a fuel, natural gas has significant environmental benefits to
offer. It is important to know that its continued use is assured.

The known reserves of natural gas have been increasing world-wide for many years, even at a time when the emphasis was actually on exploration for oil rather than for gas. When the ratio of reserves of natural gas to production is examined (Fig. 4.9), it is clear that there is an adequate supply, and the ratio is expected to remain high even with increased usage. However, it is not a readily renewable resource, and the danger of over-emphasis of short-term advantages as against long-term realities must be avoided. For example, there are enormous reserves of coal and it is inconceivable that these will not ultimately be brought into greater use. There are already environmentally acceptable and proven gasification methods in which carbon dioxide (and sulphur) can be removed in a concentrated form, although the fixing or long-term storage of carbon dioxide poses a greater challenge.

Can this gas resource be brought to the European market place? The means are already to hand. There is an integrated

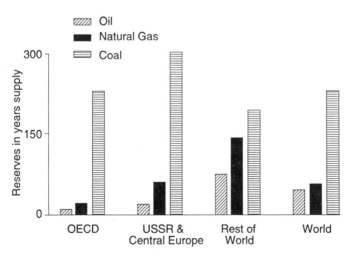

Fig. 4.9 Ratio of fuel reserves to production at the end of 1990. (From BP Annual Statistics, 1990.)

high-pressure pipeline supply system to the European market from the North Sea, Russia and the other independent republics of the former USSR, and North Africa, to be reinforced in time perhaps from the Middle East. In addition, there are established LNG routes by sea which are expected to be expanded over the next two decades.

In the search for the perfect fuel with no environmental impact we often hear mention of the hydrogen economy. However, there remain a number of fairly intractable problems to be overcome for large-scale distribution and use of hydrogen—even when there are practical economic methods for large-scale manufacture. Hydrogen has a very high flame speed which makes it prone to explosion with great violence, and it is notoriously difficult to contain. It cannot be regarded as an ideal fuel for widespread domestic use.

In the future, new business opportunities will bring different challenges which will need to be continously addressed. Every company operates against a changing background of risks, legislative controls, and public anxieties. As public awareness of risk and environmental issue continues to increase, so will the pressure from regulatory authorities, environmentalists, and local authorities, to demonstrate that the risks posed by gas industry operations, and any effect on the environment, are acceptable. The implications of eco-labelling, the estimation of non-energy costs for different energy routes, and other developing concepts, have still to be evaluated. In its long history the gas industry has shown itself to be a survivor through the imagination, skills, and flexibility of those working within it. It will continue to contribute significant benefits to the community and maintain its rightful role in the energy mix of the future.

REFERENCES

Barty-King, H. (1984). *New flame: how gas changed the commercial, domestic and industrial life of Britain between 1813 and 1984.* Graphmitre Ltd, Tavistock.

British Gas (1991). *Environmental review*. British Gas.

Button, K. (1990). *The urban environment*, Key Environmental Issues, No. 9. British Gas.

CEST (1991). *Industry and the environment: a strategic overview*. Centre for the Exploitation of Science and Technology, London.

Department of Energy (1989). *An evaluation of energy related greenhouse gas emissions and measures to ameliorate them*, Energy Paper No. 58. HMSO.

Department of the Environment (1990). *Digest of environmental protection and water statistics*, No. 13. HMSO.

Elliot, D. (1990). *Renewable energy*, Key Environmental Issues, No. 13. British Gas.

Houghton, J. T., Jenkins, G. J., and Ephraums, J. J. (eds) (1990). *Climate change—the IPCC scientific assessment*. Cambridge University Press.

McCleish, E. (1990). *Environmental education—the vital link*, Key Environmental Issues, No. 14. British Gas.

Pearce, D. (1991). *Corporate responsibility and the environment*. British Gas.

Thurlow, G. (1990). *Technological responses to the greenhouse effect*. Elsevier Applied Science, London.

Williams, T. I. (1981). *A history of the British gas industry*. Oxford University Press.

5

Alternative energy sources: the European story

John Rae

After taking his doctorate at the University of Glasgow and undertaking further research there, Dr John Rae pursued a distinguished academic career in physics, both as a lecturer and a researcher, at the University of Texas, at the Université Libre in Brussels, and at Queen Mary College, London until 1974. He then moved to the Atomic Energy Authority (AEA) at Harwell as an Industrial Fellow; in 1976 he was appointed Leader of the AEA's Theory of Fluids Group. Nine years later he moved to Whitehall to take up the appointment of Chief Scientist at the Department of Energy, as one of the successors to another Linacre Lecturer, Sir Hermann Bondi. Dr Rae was simultaneously a Member of both the Natural Environment Research Council and the Science and Engineering Research Council. In 1990, he returned to Harwell to take up his present position as Chief Executive of AEA Environment and Energy, a division of AEA Technology.

INTRODUCTION: THE EUROPEAN REVOLUTION

Recent years have seen a succession of great events change the face of Europe. The collapse of the Berlin Wall, the toppling of a succession of authoritarian regimes, and the break-up of the Soviet Union are just some of the episodes that have marked the movement towards democracy and the free market philosophy now gaining momentum in Eastern Europe. The changes that Eastern Europe has witnessed can be regarded as nothing less than revolutionary.

These changes, though, form part of a deeper European revolution that has seen the character of this continent alter profoundly since the Second World War. Greater economic interdependence, the superseding of national barriers by industry and commerce, and the forging of ever closer ties between nations, especially through the European Community, have been of fundamental significance. The overriding aim of these developments has been to increase Europe's economic prosperity.

All the signs are that the coming years and decades will witness continued economic growth in Europe, primarily through the development of a Single Market, which more and more countries are eager to join. Indeed, once the initial problems associated with switching to a capitalist system have been overcome, the emerging democracies of Eastern Europe will in all probability enjoy excellent rates of economic growth. And if the present tendency towards the fragmentation of large states such as the USSR and Yugoslavia continues, the Europe of the future could be characterized by a proliferation of nations whose economies are expanding rapidly.

ENERGY AND THE ENVIRONMENT: EUROPE'S DILEMMA

Energy is the life-blood of any economy. Without a reliable, competitive supply, industry and commerce cannot function. As economies expand so does their thirst for energy. A future of economic growth in Europe will inevitably mean that Europe's demand for energy rises to unprecedented levels.

In addition to Europe's Economic Revolution, however, recent years have witnessed what might be termed a Green Revolution. This has involved a recognition that the environment is an issue of the utmost international importance. The environmental price that has to be paid for some of our activities—in industry, transport, agriculture, and so on—is slowly dawning on us. One has only to think of the monstrous scale of the environmental vandalism that has been wrought in Eastern Europe, the results of which are only now beginning to come to light. A consensus has emerged across Europe that, in many cases, the price we have paid has been too high.

One activity with far-reaching implications for the environment is the generation of electrical power. The large-scale, continent-wide environmental effects of power generation, such as the emission of carbon dioxide and the pollutants that cause acid rain, are well documented. But these effects represent only the tip of an iceberg that also comprises many, more localized environmental issues, such as the siting of power plant and the transportation of waste. Some of these issues have been raised in earlier presentations in this Linacre Lecture series.

The problem, then, is clear cut. Europe will need more energy as its economies develop, but this energy will have to be supplied at a lower environmental cost. Judging by past experience, these are potentially contradictory objectives, but they will have to be reconciled. As the Brundtland Report stated (World Commission on Environment and Development 1987, p. 38): 'We are now just beginning to realise that we must find an alternative to our misplaced belief that there is a choice between the economy and the environment.'

It is one thing to call for a Europe where energy and the environment go hand in hand, quite another to bring it about. Europe is a heavily industrialized and energy-intensive continent. And as already outlined, further growth in demand is certain. The progress being made towards using energy more efficiently will not alter that fact. Furthermore, other considerations besides the environment have to be borne in mind when looking at options for meeting rising energy demand. Security and diversity of supply are matters of the utmost importance. The Gulf War of 1991 and the events that led up to it stirred anxieties over oil supplies and were a timely reminder, if such a reminder was necessary, that over-reliance on energy imports leaves nations vulnerable to political developments beyond their control. Ideally, then, Europe should be as self-sufficient in energy as possible, relying on indigenous sources wherever feasible, while giving due weight to factors such as the environment.

Can all these criteria be met by those energy sources on which Europe has relied in the past—coal, oil, gas, and nuclear power? The answer would seem to be 'no'.

THE RENEWABLE ENERGY OPTION

Could renewable energy play a part in Europe's energy future? At first sight the renewables meet the necessary criteria: they are comparatively benign from an environmental standpoint (e.g. in terms of carbon dioxide and acid rain emissions); inexhaustible, for all practical purposes; indigenous and therefore secure. Europe's renewable resources are many and varied, with different forms of renewable energy prevalent in different regions. There is a large biofuels resource almost everywhere; solar energy, too, can be harnessed Europe-wide, though best of all in Mediterranean countries; the most favourable wind regime is to be found in the north-west; potential for hydro power exists in most countries, while the best tidal and wave resources are to be found along the Atlantic seaboard.

Work has already started on harnessing these resources. Denmark has over 300 megawatts (MW) of installed wind power capacity, with 2000 wind turbines and several wind farms in place; The Netherlands and Germany each have over 50 MW of installed wind capacity; Sweden is working on the exploitation of offshore wind energy. Work to investigate the potential of tidal energy has been pursued in the UK, France, and the former USSR, with prototypes built in the latter two countries. The French prototype barrage at La Rance (240 MW) is the largest in the world. The exploitation of the heat from geothermal aquifers is well established in France, Iceland, Hungary, and Italy. France has 66 district-heating schemes based on this source of energy, saving 200 000 tonnes of oil a year; Iceland has 40.

Yet, when put in perspective of energy supplies as a whole, the renewables have so far failed to make a substantial impact. Today, renewable energy accounts for only 6–7 per cent of the European Community's energy supplies, with broadly the same picture applying to Europe as a whole. It is certainly true that some countries derive a vast majority of their electricity from renewable sources: e.g. Iceland 100 per cent, Norway 100 per cent, Austria 74 per cent, Switzerland 57 per cent. But these contributions do not represent the fruits of recent development work. Most of the renewable energy currently being exploited in Europe (i.e. 4–5 per cent of the 6–7 per cent) is in the form of large-scale hydro power. This technology dates back to the nineteenth century and constituted many nations' first source of electricity. It now provides the Community with 12 per cent of its electricity and even the comparatively mountainless UK derives 2 per cent of its electricity from large-scale hydro schemes.

Other technologies are either some way from the market-place (e.g. tidal and wave power) or limited in their deployment in practice (e.g. wind, biofuels, and photovoltaics). This is a paradox of the renewables. The scale of the debate that surrounds them is out of all proportion to the energy they presently deliver.

Bearing in mind the widespread nature of the resource, though, how great a contribution could the renewables make? In terms of technical potential, the renewables could meet all of Europe's energy needs. Even after various constraints have been taken into account (e.g. unavailability of land) they could provide a significant fraction of Europe's energy in the twenty-first century. Individual members of the European Community envisage increasing the use of renewable energy to meet 15–20 per cent of their needs by the early twenty-first century.

Let us look at which are the most promising technologies for Europe. Potentially, wind power alone could meet all of the European Community's electricity needs. Studies have shown that 15 per cent of the European Community's total electricity could be generated from the wind at available onshore sites and accepted by current networks. Many countries have set themselves targets of several hundreds of megawatts of wind power capacity installed by the year 2000. Biofuels in Europe are chiefly waste and fuel wood. The exploitation of landfill gas is well advanced in the UK, while elsewhere in Europe energy recovery from waste incineration is widely practised. The use of fuel wood is well established in Sweden and Finland, and the Community currently uses 20 million tonnes of fuel wood a year. But a much bigger wood resource is available. If the 20 million hectares of land in food overproduction were turned over to the growing of arable coppice, the fuel wood produced could provide 20 per cent of Europe's energy. Photovoltaics could supply a few per cent of Europe's energy needs. At the moment, its market in Europe is very modest, but interest in this technology is growing markedly. Thousands of remote houses in Italy and Spain have recently been electrified thanks to the installation of solar power systems. Germany has a '1000 roofs' programme to install grid-connected systems on private houses, and the development of centralized photovoltaic electricity generation plant is being pursued in Italy and elsewhere.

It is not only the 'major' European countries that have the potential to benefit substantially from the renewables. Greece, for example, has considerable wind, solar, geothermal, and biofuels resources. Ireland has potential for small-scale hydro, wind, biofuels, and wave power.

IS THE UK LAGGING BEHIND EUROPE?

Renewables could make an important contribution to the European energy market but have yet to do so. It is not just in the UK that they have still to realize their potential. The UK, none the less, is frequently chastised by its domestic press and by environmental pressure groups over its limited use of renewables. Accusations of 'foot-dragging' are commonplace.

Renewable energy currently accounts for about 2 per cent of the UK's electricity. On the face of it, this does not compare well with some other European countries, but there are factors to be taken into account. The UK is well furnished with indigenous fossil fuels (North Sea gas and oil, and comparatively rich coal reserves). It was the lack of such fuels that was perhaps the major factor in prompting 'progressive' renewables-developing countries such as Denmark and The Netherlands to develop their renewables resources at a slightly faster rate. The UK has comparatively few communities so remote that they cannot be connected to the electricity grid, unlike more mountainous countries and countries with more scattered populations. In fact, in the UK, remote communities are already benefiting from the exploitation of the renewables: several islands have their own wind turbines, while Islay has a prototype shoreline wave device. The UK does not have the community-based culture of, for instance, Denmark, which has given rise to the Danish biofuels-powered district heating systems, often fired by straw. These systems produce hot water centrally for distribution to homes, schools, factories, and so on. British 'individualism' has meant that we prefer to provide our own hot water from our own individual boilers. In some other countries, renewables are already established, as a result of nature or history, or both. In Sweden, for example, biofuels (principally wood) meet one-seventh of energy needs. But Sweden has a well-established, world-class forestry industry.

Despite these factors, 'new' (i.e. non-traditional) renewables are not making a much greater contribution in other countries than in the UK. The UK spends about the same on renewable energy Research Development and Demonstration (RD&D) as Germany, Italy, and The Netherlands. Prompted like its counterparts in the rest of Europe by the 1970s world oil crisis, the UK Department of Energy set up its Renewable Energy RD&D Programme in 1974. Since then, the Government has spent £180 million on the programme, with a record spend of £24 million in 1991–2. Largely as a result of this programme, the UK is ahead of the European field in some areas:

- landfill gas—worldwide, only the USA has more sites being exploited commercially than the UK;
- tidal power—current interest in this technology is virtually confined to the UK, with ongoing studies taking place of potential barrages on the Severn, Mersey, and a number of smaller estuaries;
- wave power—the UK is well to the fore; only Norway can compete as a leading developer;

- geothermal 'hot dry rock' research—this has been the domain of the UK, Germany, and France; in the UK a programme has been pursued at Rosemanowes in Cornwall since the 1970s and it will soon become part of a European Collaboration.

The UK, then, is ahead in some renewables, but behind in others, often for very good reasons. In particular, the UK's record of research and development is good. The UK is a little behind some countries, though, when it comes to market penetration by some promising technologies (e.g. wind). But overall, the UK's record is comparable with that of others.

Where the UK may be open to criticism is in the area of targets. While it is commonly stated that renewables could provide up to 20 per cent of the UK's electricity by 2025, the current short-term goal—1000 MW of new renewable installed capacity by the year 2000—does not appear overly ambitious, particularly when we remember that the UK has the best wind resource in Europe. The Netherlands, for example, aims to have installed 1000 MW of wind power alone by the year 2000, part of its overall objective to meet 10 per cent of primary energy demand from renewables by 2010 (to be achieved by waste incineration and solar power, in addition to wind).

But what is more important than the goals themselves is the extent to which they are realistic. There is no point setting ambitious targets if they cannot be realized, and a number of barriers have become apparent that might hamper the extent to which renewables fulfil their potential and the rate at which they do so. These problems have little to do with technical viability; the barriers are primarily non-technical in character. Nor are they unique to the UK. They are to be found Europe-wide.

BARRIERS TO INCREASING RENEWABLES

One set of barriers comes from the economics of renewables. The European Commission's report *Energy for a new century* by the 'Groupe des Sages' in 1990 identified an overall need to reduce unit costs in order to make renewables more competitive with conventional power supply options. To some extent, such a reduction might be facilitated by further research and development. But mass production of wind turbines or photovoltaic cells, for example, would make the greatest contribution to cost reductions. Those cost reductions would then create a mass market for renewables. But industry will not countenance mass production unless a mass market already exists. This leaves the renewables in a classic chicken-and-egg situation, with market forces rebelling against their wider uptake.

Some renewables, though, are already cost competitive or are on the verge of becoming so (e.g. landfill gas, waste combustion, wind power). The more problematic technologies, economically speaking, tend to be those which involve a high level of up-front capital expenditure and long payback periods (e.g. tidal power), where projects only look attractive if they are given long-term contracts and assessed at low discount rates. The inclusion of external costs such as environmental impact into the price of power generated might well make the renewables look more competitive compared with conventional generation methods.

The next set of barriers arises from environmental issues. Development of wind farms in the UK has prompted some opposition on localized environmental grounds. Above all, opponents cite visual intrusion onto the landscape, the noise that wind turbines are claimed to make, and the loss of landscape value. European wind power development has met with the same kind of opposition. Similarly, opposition has arisen across Europe to waste incineration projects because of fears of emissions. New large-scale hydro schemes, too, provoke disquiet, largely owing to the loss of land they involve. In Sweden at present, hydro is lumped together with nuclear power as an environmentally unfriendly means of generating power. The new environmental awareness of the modern day, then, has proved a double-edged sword. On the one hand it has stimulated interest in renewables in general, on the other it has given rise to opposition to individual renewables projects.

There is also a class of institutional barriers which stem from the fact that society ('the system') is not geared to exploiting renewables. Wind power is a good example. Potentially, wind could meet all of the European Community's electricity demands. However, planning restrictions and the problems of integrating wind power into existing electricity supply infrastructures combine with physical restraints (e.g. the fact that wind turbines cannot be placed in towns or on railways) to reduce exploitable wind power to 10 per cent of the European Community's demand. Development of suitable storage and conversion systems might increase the potential. At the moment, though, wind power is less developed in Europe than in the USA, which has 1500 MW of installed capacity. In The Netherlands, where the potential for onshore wind power amounts to 8700 MW, only 1000–1600 MW of that amount could be absorbed by the grid without operational penalties. Other restrictions are imposed by Dutch policies on noise abatement. Even though the public belief that wind turbines are noisy is a myth, the strict Dutch regulations enforcing the maintenance of 'silent' areas limits the sites available for wind farms. And this in a country so used to windmills! To sum up, the biggest single hurdle facing the development of wind power in Europe is the obtaining of planning consent. It will not prove easy to find enough environmentally acceptable sites at favourable high wind-speed locations.

The story for other technologies is similar, particularly with regard to the absence of an infrastructure geared to exploiting them. Incineration of wastes such as straw, chicken litter, and scrap tyres is a case in point. Europe generally lacks the infrastructure that would make possible the supply of those wastes to incinerators in the quantities necessary to make schemes commercially viable.

Despite the broad public consensus that renewable energy is to be welcomed and is preferable to other forms of power generation, innate conservatism and a degree of ignorance about the realities of renewable energy combine to generate scepticism and, sometimes, hostility towards renewables technologies and individual projects. This is true of both the public and industry.

Most of these non-technical barriers are common to all energy sources, indeed to developments of all kinds. NIMBY (not in my back yard) and BANANA (build absolutely nothing anywhere near anyone) attitudes are not confined to the UK nor to any single technological development. Motorways, rail-links, and 'conventional' power stations face the same kind of environmental and planning problems that confront the renewables. The renewables do not offer a short-cut around these problems.

BREAKING DOWN THE BARRIERS

Now it is time to turn to the third part of the equation. Will the UK continue to keep pace with other countries or will it be left behind? That is to say, will the barriers be surmounted more quickly or more slowly in the UK than elsewhere? It is one thing to identify barriers to exploitation, another to find ways of overcoming them. A number of approaches and initiatives have been and are being devised across Europe.

It is generally recognized throughout Europe that market forces in themselves are not enough to ensure an expansive future for renewables. If the environmental and other advantages of renewable energy are to be capitalized on, some form of government intervention is needed (besides subsidizing research and development), in tandem with action by industry and the community. Many countries have introduced incentives and/or regulations with a view to developing the markets for renewable energy. These vary from country to country.

Financial support comes in various forms: subsidies for users and manufacturers; tax credits; technical assistance for developers; guaranteed prices for energy produced. The German central and regional Governments provide subsidies of up to 70 per cent for the installation of photovoltaics on private houses. This scheme has met with a favourable response from the public. The Greek Government introduced financial incentives to kick-start the solar market between 1978 and 1982. These incentives, combined with the prevalence of flat roofs in Greece, ensured that the market developed successfully. In Germany, The Netherlands,

and Denmark, the Government has established financial incentives for buyers and/or manufacturers of wind turbines. The UK
has the Non-Fossil Fuel Obligation (NFFO), whereby regional
electricity companies in England and Wales are obliged to derive a
specified amount of their electricity at premium prices from
renewables generators. Originally devised to protect nuclear
power after privatization, the NFFO has nevertheless proved successful in stimulating a market for renewables, with nearly 200
projects in a range of technology categories (wind, waste, landfill
gas, small-scale hydro, etc.) already accorded NFFO status.

Legislation has been tried in many ways: the following are some
examples. The NFFO in the UK has set up a protected market for
renewables, to help them penetrate the electricity supply industry.
The UK Government has introduced a variety of other measures
too, designed to allow the renewables to compete with conventional, large-scale methods of power generation on a 'level
playing-field'. These have included the reform of the local authority rating system, which had discriminated against small-scale,
independent electricity producers in favour of the Central Electricity Generating Board, and the removal of water abstraction
charges imposed by the water industry on developers of small-
scale hydro schemes. In Italy, a law of 1987 encouraged the
exploitation of municipal waste as a source of power by requiring
the separation of recyclable components from the rest of the waste
at source. Perhaps the most spectacular example of a legislative
initiative was the Swedish Parliament's decision in the early 1980s
that Sweden should in future rely on preferably lasting, indigenous
energy sources with the least environmental impact, thereby ruling
out nuclear power and large-scale hydro power and limiting the
use of fossil fuels. This has hastened the rate of renewables
research and development in Sweden. In Denmark, the advanced
status of straw as an alternative fuel has been engineered by the
imposition of a punitive tax on the domestic use of fossil fuels,
making those fuels two or three times more expensive than they
would otherwise be. A carbon tax on conventional energy—
designed to combat carbon dioxide emissions—is being discussed

by a number of countries besides Denmark, as well as by the European Commission.

As well as individual governments' efforts, international bodies such as the International Energy Agency and the European Commission (CEC) are acting to stimulate the take-up of renewable technologies. The CEC, which since 1975 has contributed financially to over 1000 renewable energy research and development projects, is providing financial assistance under the 'Valoren' programme for wind farm projects in Ireland, Spain, Portugal, Italy, and Greece, while the WEGA (Wind Energy Large Machines) collaborative project, part-funded by the CEC, aims to demonstrate megawatt-size wind machines for electricity generation. The CEC's non-nuclear research and development programme JOULE, and its THERMIE, programme are also having an impact, in photovoltaics for example. Besides providing project support for work on innovative technologies, THERMIE places particular emphasis on the co-ordination of work undertaken in different member states and on the effective dissemination of project results. Such dissemination is intended to enhance industrial awareness and increase industrial participation in the development of renewable energy technologies.

Of equal importance to the raising of interest within industry is the forging of support for renewables among the general public. The NIMBY/BANANA syndrome has to be tackled; this can only be achieved by educating the public. For most lay people, the renewables are an unknown quantity and it is the fear of the unknown that often lies at the heart of opposition. People must be fully informed about the difficult choices that have to be made in the field of energy supply, the pros and cons of renewables and the reality of living near a wind farm, for instance. Studies into public attitudes towards renewables carried out in the UK and The Netherlands have highlighted the importance of keeping the public fully informed of plans for wind farms and so on well in advance of official announcements and developments. The NIMBY syndrome is notably absent in those cases where 'ordinary' people and local communities are involved in the planning process.

In order to break down the barriers limiting the exploitation of renewables, then, there are a number of possible approaches. Inevitably, different approaches will suit different technologies in different places at different times, so flexible intervention by government will be needed, though this will not necessarily involve a financial commitment in every case. So which countries have the right approach today? Whose initiatives will prove the most successful in promoting the use of the renewables? As yet, it is impossible to say. But by the year 2000, we will know who had the best policies and the most appropriate targets.

WHAT NEXT?

The relatively small contribution that the renewables are making in the UK is not due to some quirky British parochialism that flies in the face of a European trend. Barriers exist throughout Europe and arise not so much as a result of national characteristics but rather as a result of human nature. It takes time for people to embrace novelty. The renewables may constitute a family of alternative technologies, but they do not offer an alternative route around deep-seated human attitudes. The exact contribution made by renewable energy in the future will lie somewhere between its technical potential and the amount currently being employed. The choices and attitudes of people—private individuals as well as industry and governments—will determine the precise size of its contribution.

The future, therefore, is in everyone's hands. But what would that future look like if Europe chooses not to embrace the renewables? It is a future where energy demands prove ever harder to meet, and where the prospect of economies contracting and living standards falling is traded against ever more acute environmental problems. It would be wrong to say that, alone, the renewables can prevent those nightmares becoming real. In that sense, they do not represent alternatives to conventional energy sources, but rather sources complementary and additional to those we rely on today. Nevertheless, they have the capacity to make a vital contribution.

REFERENCES

Groupe des Sages (1990). *Energy for a new century—The European perspective*. European Commission, Brussels.

World Commission on Environment and Development (1987). *Our common future* ('The Bruntland report'). Oxford University Press.

6

Energy generation in Central and Eastern Europe: the environmental problem

Peter Hardi

Dr Peter Hardi, a Hungarian political scientist, took his first and higher degrees at the Eötvös Lorand University in Budapest and was subsequently awarded an Academic Degree in Political Science by the Hungarian Academy. He has taught for twenty years at the Budapest University of Economics (formerly the Karl Marx University) and during the 1980s visited the United States both as a Visiting Professor at Yale and as a Research Associate at the Institute of East/West Studies in New York. In 1988, following his return to Budapest on being appointed Director of the Hungarian Institute of International Affairs, Peter Hardi became actively involved in the mounting controversy surrounding the construction of a major hydroelectric dam at Nagymáros, on the Danube Bend north of Budapest. This mega-project, sponsored and financed by both the Hungarian and Czechoslovak Governments, came under increasingly bitter attack from Hungarian liberals and environmentalists as soon as the process of political reform, which culminated in the ousting of the Communist regime, made free and uninhibited comment possible. Dr Hardi was appointed by the Hungarian Government to head a panel of international experts to report to Parliament on the environmental and economic implications of continuing the Nagymáros project, which had by then moved to the centre of national political debate. The 'Hardi Report' found decisively against continuation and was largely responsible for the Hungarian Parliament's decision, in 1989, to call a halt to construction at Nagymáros. Dr Hardi was subsequently appointed to his present position of Director of the Regional Environmental Centre for Central and Eastern Europe in Budapest; he is also a member of the Environmental Advisory Council of the European Bank for Reconstruction and Development.

Experts usually agree that a switch from heavy industry to more service-oriented industries, changes in energy generation and energy use, as well as changes in agricultural methods, are all critical elements in preventing the further degradation of the Central and Eastern European environment and in improving it in the longer run. It is a commonplace that the restructuring of the entire economy in each of the Newly Democratizing Countries (NDCs) of the region is a *sine qua non* of any sound and effective environmental policy (Schreiber and Weissenburge 1991).

The optimistic approach sees an extraordinary possibility for improvement through political change, which opens up avenues for genuine structural change in the entire economic field. According to this argument, after the collapse of the communist system the previous economic policy will also disappear, and the transition to a market economy will make economic restructuring inevitable. As a consequence, the restructuring can be oriented and influenced to create less polluting, more environmentally friendly industries, and to close down the most ineffective and polluting plants. The problem is that both the transition and the restructuring are more painful and slower processes than anybody anticipated.

The basic *political* dilemma of the new governments is that they won elections not only by a promise of a democratic and free society, but also of a better future, and an improvement in overall standards of living. The West is attractive to NDCs not only because it has an open and democratic society but also because it offers a high economic standard of living. Governments of NDCs have an elector obligation to meet this expectation; moreover, they do not feel able (in other words, they are too weak) to introduce highly unpopular measures which, in the short run, are contrary to electoral expectations. So the basic political reason for not introducing sweeping measures for economic transformation as soon as possible is fear of the sociopolitical consequences.

The basic *economic* reason for postponing sweeping reforms is the extent of the immediate economic problems these governments have to face. Their short-term and, in many cases, medium-term preoccupation is a desperate struggle for economic survival.

In other words, the hands of the new governments seem to be tied, and they are deprived of sufficient freedom of action. The extent of their limitation is a heritage of the past which had not been anticipated.

A comparison between the NDCs and the former GDR demonstrates that lack of resources really limits the scope of possible action in completing economic restructuring. But even in the eastern part of Germany, which has all the resources of western Germany at its disposal and is legally integrated into the West by a political act, the process of transformation will take much more time than anticipated. The case of Germany provides additional proof that even if governments could overcome their political fears and had access to the necessary financial means, the process of economic transition is still going to be lengthy and complicated, with a delay between decision-making and actual implementation (Calvo and Frenkel 1991; Comisso 1991; Milenkovitch 1991).

All economic difficulties which hinder the creation and implementation of a proper and effective environmental policy have to be considered in the context of a major structural change: the transition from a centrally planned economy to a market economy. This process is not simply a domestic affair; it is closely related to the extinction of the Council for Mutual Economic Assistance and the collapse of the eastern bloc market. Beyond this fundamental issue, environmental policy-making has to face specific structural economic problems which determine the present structure of industry and the energy sector. In addition to the transition-related and structural problems, there is the issue of scarcity of resources (both financial and technical), international debt, and also the foreseeable problems of structural changes and transitions in employment policy.

The greatest structural problem is the heritage of an over-large, in most cases completely obsolete, extremely energy-inefficient heavy industry, the operation of which is usually deeply in the red. Restructuring requires not simply privatization and the closure of inefficient factories, but a wholesale modernization programme. A backward industrial structure, together with the irresponsible policies of planned economies, are mainly responsible for ecological destruction in the region. Yet, beyond privatization plans, the governments have no comprehensive industrial or modernization policies (Milenkovitch 1991, pp. 159–60). It was not foreseen what kind of problems would be generated by the collapse of the arms industry (hitting Bulgaria and Czechoslovakia particularly strongly), by the radical decline in Soviet oil deliveries, or by the disintegration of the Comecon market. Given these difficulties, further industrial cutbacks and closures are much more difficult to implement than was previously envisaged.

During the transition period, it was also hoped that the products of communist ideology such as huge steel-mills and aluminium works could be eliminated. These plants were created by an underlying philosophy according to which socialism had to be based on heavy industry and on a working class closely related to that industry. In many cases the creation of huge heavy industry installations was based on the intention to replace local intelligentsia or peasantry with heavy industry workers (for example the steel-works at Nowa Huta near Krakow in Poland, or the aluminium works at Ziar nad Hronom in Slovakia). According to conventional wisdom these highly polluting symbols of communist industrialization should have been among the very first plants to close down during industrial restructuring. Yet the legacy of the past seems to be stronger than anticipated.

Because of the lack of comprehensive industrial policies and transition plans, and the pressure of immediate liquidity problems, the influence of large, monopolistic industrial structures, and the bargaining power of their lobbies, has not yet disappeared.[1] At the same time, social tensions aggravate the situation. Even issues of nationality play a role in maintaining highly polluting industrial structures.[2]

Another structural problem is posed by the energy sector itself. In the ex-communist countries energy needs for the production of the equivalent of US$ 1 of national product is among the highest, while energy efficiency is among the lowest in the world (IEA 1990; Levine *et al.* 1991). The industrial sector in the region consumes approximately twice as much energy to produce US$ 1 of product as industry in the United States (Kolar and Chandler 1990, p. 8). At the same time, in most countries of the region the main sources of energy are highly polluting brown coal and low-quality soft coal or lignite (French 1990, p. 11). Thus an efficient and very polluting energy sector provides the basis of an obsolete and extensively polluting heavy industry. Though references to energy efficiency and conservation appear in the programmes of the new governments, long-term energy programmes nevertheless include the creation of huge new power plants.

Despite the fact that actual energy consumption declined significantly in 1990 owing to higher energy prices and industrial recession,[3] and despite the recommendations of all relevant international studies demonstrating huge reserves in energy savings (IEA 1991; Levine *et al.* 1991; Schreiber and Weissenburger 1991), and even showing significant energy export potentials (Radetski 1990), and despite the fact that energy demand in Central and Eastern Europe can be held to current levels until the year 2025 (Kolar and Chandler 1990, p. 18), the alternative presented by most governments is not between the present trends and a serious reduction in energy consumption, but between different methods of increasing energy output. In Hungary, for example, the alternatives presented by the Government are a huge new lignite-based power station or the extension (doubling) of the existing nuclear power station.

The underlying logic is almost identical with that of the communist regime: future industrial recovery and expansion will demand more and more energy. The lack of understanding among political leaders for the need for a sustainable development policy—the only present exception is the Polish environmental administration with its emphasis on sustainable development (MEP 1991)—and the vested interests of technocrats in the energy sector (who are sometimes rivals for resources but do not differ in approach) must also be considered legacies of the past, making a shift towards a new, environmentally oriented, conservationist energy policy more difficult.

The old logic of energy dependence is also evident in the well-known controversy between the Hungarian and Slovak Governments over a hydroelectric power station, the Gabcikovo Nagymáros barrage system (GNB). From my perspective, it is an extremely relevant example of the hidden heritage of the past.

The GNB is a symbol of communist industrial mismanagement and secretive decision-making. It mobilized environmentalists who considered the GNB a major threat to the environment and generated the first major street demonstration (of about 30 000 people) since 1956 in still-communist Hungary. It was a major policy issue on which the communist leadership was defeated in the fall of 1989 in Hungary (Hardi *et al.* 1989). Conventional wisdom had predicted an immediate halt to the project, which was a symbol of gigantomanic communist industrial policy, totally unresponsive to environmental considerations and dismissive of all public opposition. Yet this has not happened in Slovakia. Needs for new sources of energy supply are greater than the fear of ecological risks; the leadership considers an increase in energy production as an outstanding priority and as a part of a more independent national economy.

The links between large natural deposits of hard coal and lignite, a wasteful and rigid economic and ideological system, an outdated and inefficient heavy industrial infrastructure, and chronic and sometimes acute atmospheric pollution have been easy to identify. At the heart of the matter lies the way in which energy has been acquired and utilized in the East–Central European countries—this is the characterization of the situation by a Western analyst (Russell 1990, pp. 23–8). Beyond summarizing qualitative statements and highlighting determinant relationships, this paper does not offer comprehensive data on Central and East European energy indicators, sector distribution patterns, efficiency and pollution data, etc. These can be found in a comprehensive survey carried out by the Organization for Economic Co-operation and Development and the International Energy Agency (OECD/IEA 1990).

There is quite a range of different fuel mixes in the energy consumption balances of the East–Central European countries. The share provided by coal reaches 78 per cent in Poland (making it one of the most coal-dependent countries in the world),

55 per cent in Czechoslovakia, but only 21 per cent in Hungary. Oil supplies 14 per cent in Poland, 21 per cent in Czechoslovakia, and 28 per cent in Hungary; natural gas meets only 8 per cent in Poland, 13 per cent in Czechoslovakia, and 29 per cent in Hungary; primary electricity (hydro and nuclear) meets only 2 per cent in Poland, 9 per cent in Czechoslovakia, but 21 per cent in Hungary (in the latter case, mostly nuclear).[4]

According to the World Bank's environmental report on Poland, the country's energy policy has been mainly based on coal. Her domestic energy supply and her energy export are dominated by coal. Energy consumption depends primarily on coal converted to electricity, and is organized on the basis of district heating. At the same time, a third of the total energy supply is either lost in the delivery systems or is absorbed in energy conversion (World Bank 1990).

The Czechoslovak economy is characterized by enormous energy demand. Consumption of energy per capita is higher than any developed country except the United States and Canada, while the per capita gross domestic product (GDP) and the overall standard of living are significantly lower. The energy sector has a highly unfavourable energy balance structure, both in the distribution of primary sources and in final energy consumption, since electricity plays a small part. The major problem, however, is that about 62 per cent of electricity is consumed by industry or production in general, leaving a little more than one third for residential consumption (in more developed countries the proportions are just the opposite (Lencz and Balajka 1990)).

In Hungary, coal mining activities related to the generation of energy have the worst impact on the environment. Damage to the surface is caused by strip-mining (most of the coal and lignite used for energy production is strip mined); mining also increases soil instability, leading to soil erosion; and the discharge of mine water is a considerable source of groundwater pollution (Hindrichsen and Enyedi 1990, pp. 30–2).

The situation in Bulgaria is almost identical: the major primary energy resource is very low-quality lignite, extracted mainly by open-cast mining. Damage to the surface is not made good.

Despite the improving technologies which make it possible to use extremely poor quality coal in thermal power stations, the burning of lignite produces the bulk of sulphur dioxide (SO_2), oxides of nitrogen (NO_x), and particulate pollution. Technologies to reduce these emissions are still at a very early stage of development in Bulgaria (Konstantinov 1990).

The issue of nuclear power provides additional examples of the hidden legacy of the past, albeit from a different perspective. Environmentalists usually oppose nuclear power in the countries of the region partly for safety reasons and partly because of the unresolved problem of nuclear waste. In Hungary, for example, one of the most celebrated cases which united the then opposition groups was a protest movement against a proposed low-radiation nuclear waste disposal site at Ofalu (mid-southern Hungary).

Opposition to nuclear power has increased since the Chernobyl accident and since operational problems emerged in the Kozloduy (Bulgaria) and Bohunice (Czechoslovakia) nuclear power stations (Ecoglasnost 1990; Rippon 1990; Thomas 1990; ÖÖI 1991). All the operating reactors in the ex-eastern bloc countries are of Soviet design and until now the countries concerned (Bulgaria, Czechoslovakia, and Hungary) have had arrangements with the former Soviet Union according to which the Soviets have handled high-radiation nuclear waste from these reactors. Bulgaria and Hungary are heavily dependent on nuclear power for the generation of their elecricity supply (close to 40 per cent and already more than 50 per cent respectively).

At the same time, these countries are seriously short of electricity generating capacity, and a higher reliance on fossil-fuel plants, especially those which use lignite, would significantly worsen air quality, and increase carbon dioxide (CO_2) and SO_2 emissions in a region which is already the most polluted in Europe.

Those countries which try to reduce their dependence on Russia or other parts of the former Soviet Union cannot make a simple switch from Soviet to Western technology. Although some Western companies are more than ready to participate in nuclear power station construction, they carry out aggressive marketing and try to use government pressure to improve their chances (see especially the case of Electricité de France, but Siemens, Westinghouse, and the Atomic Agency of Canada Ltd (AECL) are also involved (Woodard in press)).

Poland ruled out nuclear energy as an alternative for the next decade. The government cancelled construction work on the country's first two nuclear power units at Zarnowiec, although the first unit was around 70 per cent complete, and pressure vessels for both units had been delivered and installed (Rippon 1990, p. 61).

The nuclear programme in Czechoslovakia is also endangered. There has been severe criticism from Austria of the Jaslovske Bohunice twin reactors (type V-1) for safety reasons (the reactor is located less than 40 miles from Vienna), and after a series of Government negotiations an investigation showed concern to be justified (Österreichische Expert Commission Bohunice 1991). The Czechoslovak Government faces strong political pressure from Austria as well as from its own non-governmental organizations (NGOs) to close the plants down. Another nuclear power plant planned to be built at Temelin has already generated protest from Slovak and Hungarian NGOs.

In Hungary, on the other hand, nuclear energy seems to be the favoured option of the government. The Government is considering doubling the capacity of the country's only nuclear power plant at Paks (which already provides approximately 50 per cent of the country's electric supply). The surplus electricity produced by the power plant could be exported to neighbouring countries. The investment costs, however, would further increase the already largest per capita foreign debt in Europe by up to 20 per cent (the cost of a single 960 megawatt reactor is the equivalent of 10 per cent of Hungary's foreign debt of US$ 21 billion (Kats 1991)).

Romania had already decided to accomplish a nuclear programme with Western support during the rule of Nicolae Ceaucescu. AECL suspended construction work in 1989 owing to the extremely serious shortcomings in quality of the Romanian construction work. The present Romanian leadership, opting for nuclear energy, renewed the contract; the Canadians have full control over the construction of the five-unit reactor at Cerna-voda (a Canadian Government loan of US$ 277 million provides financial backing).

Bulgaria's six-unit nuclear power plant at Kozloduy, located on the Danube, is sometimes referred to as the nuclear time bomb of Europe (it was shut down seventeen times during its first two years of operation) (Tavrilov 1990). Bulgaria depends heavily on nuclear power for its electricity generation, and despite instruc-tion from the International Atomic Energy Agency to shut down the six units for much needed repairs and safety modifications, authorities could not fully comply and two reactors are still in operation. Bulgaria has no money either to import electricity or to pay for the renovation (an estimated cost of US$ 250 million). At the same time, also under the pressure of environmentalist protest, the government decided to stop the construction of a second six-unit nuclear power plant at Belene in mid-1991.

The conflict between those supporting development of greater nuclear generating capacity to ameliorate atmospheric pollution generated by fossil-fuel combustion and those who support alternative responses, is amply illustrated in the East–Central European context. A degree of popular opposition to nuclear energy has emerged at a level unprecedented in recent decades.

At the same time, the renewable energy sector is not even in its infancy in the region, and it will be many decades before it can begin to contribute significantly to the energy balance, even where, as in Poland, outside assistance in the development of wind-powered generators using Scandinavian technology is being received on a small scale. The modernization of existing, extremely inefficient conventional power plants in which elec-trical output is less than one-third of the thermal input, a change to gas–steam combined cycle plants, or combined heat and power production could also significantly decrease energy consumption and increase efficiency.

Another option is to improve the fuel mix by importing better quality energy carriers such as electricity and natural gas (and to cease exporting top quality domestically produced fuels). Once again, the problem is largely financial and economic, although there are also logistical and infrastructural difficulties. The infrastructural difficulties revolve around the absence of appropriate or adequate distribution networks and up-to-date, efficient end-user equipment (burners, boilers, and electric motors). The financial problems reflect partly the lack of competitiveness of most East–Central European products on the world market, but also the level of government subsidy to which private and industrial consumers have become accustomed. Hungary has already cut subsidies for energy production and industrial consumption, but still maintains subsidies for households, though rapidly increasing energy prices are a move in the right direction.

Now that the international order has changed so rapidly, the energy relationship between the successor republics of the former Soviet Union and the countries of East–Central Europe has also changed. The countries of the former USSR, furthermore, no longer have the oil export capability of the past. East–Central European countries can no longer rely on existing, let alone increased, deliveries or attractive soft-currency trading terms. The situation with respect to natural gas is rather different. Provided that operating problems connected with an ageing and increasingly leaky distribution and transmission network can be overcome, and that payment in hard currency can be organized, the former Soviet republics are uniquely placed to supply all the East–Central European countries. Yet there is reluctance in these countries to become even more dependent upon the countries of the former USSR, not so much because of strategic considerations but because of a deep scepticism about the ability of those countries to keep up supplies should sociopolitical conditions deteriorate further.

Decision-makers are trapped by an environmental and a political dilemma: how to choose between a potentially hazardous power generation technology and a polluting one, and how to reduce Eastern dependence without significantly increasing indebtedness to the West. Financing any investment in nuclear technology demands resources unavailable in the region. Yet governments are considering this option because they are unable to transcend a logic which is based on the necessity of increasing electricity output. They do not attach great weight to declining demand due to a longer recession period, to restructuring, and new market conditions, or to modernization and the switch to more energy-efficient technologies and products. It is also questionable whether such investment is the best use of Western assistance and of scarce resources (Thomas 1990, pp. 507–8; Kats 1991).

The present-day energy and industrial structure in East–Central Europe is, of course, a heritage of the communist past. Even a few years ago it seemed evident that the main precondition of growth was a stable supply of energy to the economy. Energy specialists demonstrated how the volume of energy consumed increased year by year, and forecasts indicated a continuing trend (Fleischer 1990).

This type of thinking has not disappeared as a result of a change in government and political structure, even though alternative ideas can now surface freely in a serious debate over long-term economic strategic planning. International comparison indicates that in planned economies significant reserves of energy have been accumulated. It is not more energy that is needed for economic growth—quite the contrary. Excessive investment in the energy sector draws away resources from the restructuring of industry, a restructuring which is a precondition of internationally competitive production, including a more efficient use of energy. Previously these countries, adapting themselves to Soviet energy supply, fell into the trap of 'cheap energy'; they spent a large part of their resources, foreign credits included, on expensive energy-producing investments and thus it cost them more and more to utilize the comparative 'advantage' of 'cheap' energy. They have had no breathing space to escape from this vicious circle and they still lag far behind world trends.

As already noted, in developed countries industry's share of total energy consumption is much smaller. The energy intensity of production and of new products is also less, and their service content is higher than in the East–Central European countries. Accordingly, the consumption of energy by industry in market economies is much less in relative terms, while the energy needs of households and services are relatively much higher. The lower industrial energy consumption is also due to the fact that technologies and means are much more up to date in the developed countries; they demand less energy (and material resources).The same applies for products, household supplies, appliances, cars, and other consumer goods. Above all, in the East–Central European countries the inefficiency of energy-producing, energy-transforming, and energy-transporting systems also contributes to the severe internal loss of energy (Fleischer 1990).

For any sound solution we have to keep in mind that the links between energy and environment do not start with power plants, exhausts, or nuclear waste. The central issue is not the right choice among different methods of energy production but a rethinking of the whole energy policy. The structure of production in the East–Central European countries is characterized by unusually high energy and material intensity, and by distorted factor and product prices. The old, and often obsolete, capital stock is unsuited to a modern competitive economy. Industrial restructuring and improved construction practices may therefore follow broader objectives, incorporating energy concerns as one of several elements. A move toward a greater reliance on markets, energy efficiency, and conservation will be spurred on by higher energy prices and the profit motive. Energy-hungry heavy industries are likely to prove uncompetitive under such a shift. According to a 1990 Polish case study, such an economic shift could reduce projected energy demand by nearly half over a period of forty years.

Almost all analysts and experts have emphasized that in order to achieve radical changes in the present condition of the environment, governments should follow the radical path of restructuring the entire economy, not simply the industrial and energy sectors. Here, however, governments run into serious difficulties, not only because they all lack the necessary financial and technological resources to implement a policy of radical change but also because issues like economic survival, immediate and short-term firefighting in budget allocation, in raising export revenues, and in most cases meeting a formidable international debt servicing requirement, push environmental considerations well down the priority list.

The lack of sufficient financial resources also means the postponement of clean-up activities and the lack of remedies to improve the existing situation. Governments take action only in clear cases of imminent danger or strong local protest, or when a case is highly publicized. There is no comprehensive action plan.

Until now, the rethinking of energy policy has been a manifest goal only in Poland, where at least a proclamation of the rationalization of energy management was published at the very end of 1990 and revised in 1991 (MEP 1990; 1991, pp. 5–6).

This document links the solution to the strategy of sustainable development (development of consumption and production which preserves the qualities and resources of the environment while protecting natural habitats). The main points of this strategy are as follows:

- efficiency and conservation through the full utilization of market mechanisms, including taxation policy and changes in price structure;
- a gradual shift in primary energy carriers towards those less dangerous to the environment (e.g. from coal to gas);
- improvement of technologies both in coal processing and combustion techniques as well as the installation and proper use of devices to prevent pollution.

Although the goals are rather general, they clearly designate the proper policy to follow, and they are also fully relevant to the other Central and East European countries.

The gap between the OECD region and East–Central Europe calls for accelerated action to begin to bridge the differences between them. The gap denotes a competitive disadvantage that may become increasingly unbearable when economies become open to investment and free trade. In addition, better energy efficiency would allow countries to buy time before other more capital-intensive and complex measures can be implemented on the energy-supply side, and to save scarce non-renewable physical resources as well as improve the quality of the environment. In this context, energy efficiency appears the most cost-effective way of meeting energy needs, while it is also the most cost-effective way of reducing environmental pollution.

Increased efforts to replace outmoded equipments with more efficient technologies would yield still greater returns. According to the case study mentioned above (MEP 1991), over the next fifteen years Poland could reduce energy by 40 per cent of current levels and save money: the cost of investment in efficient technologies would be less than the cost of building new power plants.

An energy-efficiency policy and strategy requires adequate legal and institutional frameworks. Obviously, energy efficiency depends on the decisions and behaviour of numerous organizations and individuals. The success of any action or programme depends on political commitment by governments and a different investment policy that would yield results for the user at a later stage (IEA 1990).

There is ample room for Western initiatives. At the government level, environmental aid programmes for East–Central Europe have blossomed in the recent past. The G-24 High Level Meeting decided in early 1991 to assist in solving the problems of the energy sector in Central and Eastern Europe (G-24 1991). The World Bank is assisting in a project launched in Poland in May 1990 to help develop energy resources and for restructuring; it offered US$ 250 million for project costs over US$ 600 million (World Bank 1991). An arm of the Nordic Investment Bank, the Nordic Environmental Financing Corporation, has approved its first environmental investment—US$ 45 million to help establish joint venture projects in Poland and Czechoslovakia to reduce air and water pollution and to improve energy efficiency.

Serious negotiations are going on with the Norwegian Government to reduce Poland's dependence on coal and replace it with Nordic natural gas. The construction of a pipeline serving Germany, and possibly Czechoslovakia as well, is also under consideration. Poland is receiving considerable support for its fuel switching programme. Its aim is to limit significantly the amount of coal used for power generation in order to reduce both CO_2 emissions (and thus the Polish contribution to global warming) and SO_2 emissions (to diminish its contribution to acidification).

The European Bank for Reconstruction and Development (EBRD) may be another important contributor. It is already taking part in a heat supply restructuring and conservation project started in June 1991 in Poland, lending US$ 50 million, complementing a World Bank loan of US$ 340 million (World Bank 1991).

The European Community has so far committed US$ 65 million for environmental programmes in Poland and Hungary, and will soon extend its commitment to Czechoslovakia. The International Energy Agency proposed a thematic review of technologies for energy efficiency and fuel switching in East–Central Europe as a tool for understanding on a microscale what to do in practical cases, and for recommending sectoral actions or changes that could be implemented at local or national level in the short- to mid-term perspective (OECD/IEA 1990).

Bilateral programmes are also flourishing. The United States Government has opened two energy efficiency centres in Eastern Europe, one in Czechoslovakia and one in Poland. The purpose of the centres is to develop much-needed technical expertise and analytical capabilities to promote a rapid transition to efficient and sustainable development and use of energy. The Commission of the European Communities has opened a similar energy efficiency centre in Budapest.

The aid offered so far will help, but the funds set aside for the environment are very small compared with the total package of economic assistance. Thus, it is important that environmental considerations form an integral component of all aid and trade deliberations. Lending for isolated environmental schemes while giving larger amounts of funding for damaging industrial, transport, and especially energy projects is equivalent to taking a small step forwards and a giant step backwards.

NOTES

1. Cf. the conclusions drawn in Borensztein's and Kumar's (1991) analysis of privatization procedures in Central and Eastern Europe.
2. Cf. the case of the aluminium works at Ziar nad Hronom in Slovakia. The works were established in a rural agricultural area with no sources of raw materials nearby. The works, based on imported Yugoslav bauxite and a purpose-built soft coal power station, are among the main sources of pollution in Czechoslovakia. Yet after the revolution local workers are demanding the expansion of the works, which employs nearly 8000 people, because they see no possibilities for changing their way of life. Nationalism also plays a role in their demands, with an emphasis on Slovak independence and self-reliance (NGO 1991).
3. Energy consumption decreased by 6.2 per cent in Hungary alone (CSO 1991).
4. The comparative data in international, UN, and national statistics, and in PlanEcon publications are available only for 1989, but the ratio had not changed in 1990 (PlanEcon 1990; Russell 1990, pp. 8, 9, 23).

REFERENCES

Borensztein, E. and Kumar, M. S. (1991). Proposals for privatization in Eastern Europe. In *Staff Papers*, **38** (2), 300–26. International Monetary Fund.

Calvo, G. A. and Frenkel, J. A. (1991). From centrally planned to market economy: the road from CPE to PCPE. In *Staff Papers*, **38** (2), 268–99. International Monetary Fund.

Comisso, E. (1991). Political coalitions, economic choices. *Journal of International Affairs*, **45** (1), 1–29.

CSO (1991). *Publications of the Hungarian Central Statistical Office*, April 1991. CSO, Budapest.

Ecoglasnost (1990). *Arguments of Ecoglasnost against building a nuclear power station near the towns of Belene and Svishtov*. Ecoglasnost, Sofia.

Fleischer, T. (1990). *Energy, technology, economy, environment, society*. ISTER Publications, Budapest.

French, H. F. (1990). Green revolutions: environmental reconstruction in Eastern Europe and the Soviet Union. *Worldwatch Paper*, **99**, November 1990.

G-24 (1991). G-24 assistance in the energy sector over the medium and long term, *XVII/PN/057*, 1 February 1991. G-24, Paris.

Gavrilov, V. (1990). Environmental debate creates serious problems for government. *Report on Eastern Europe*, 25 May 1990. Radio Free Europe, Munich.

Hardi, P., Vargha, J., and Fleischer, T. (1989). *The Hardi report*, Presented to the Council of Ministers and the Hungarian Parliament. Hungarian Institute of International Affairs, Budapest.

Hinrichsen, D. and Enyedi, Gy. (1990). *The state of the Hungarian environment*. Hungarian Academy of Sciences, Budapest.

IEA (1990). Closing remarks, presented by S. I. Garriba. Informal planning workshop on specific International Energy Agency activities to address energy-efficient technology and environmental problems in Eastern European countries, Budapest, November 1990.

IEA (1991). *The energy policies of Hungary*, July 1991. Survey of the International Energy Agency, Budapest.

Kats, G. H. (1991). *Hungary at the crossroads: energy efficiency or nuclear power?* Invited report to the Environment Committee of the Hungarian Parliament, May 1991.

Kolar, S. and Chandler, W. U. (1990). *Energy and energy conservation in Eastern Europe: two scenarios for the future*. Paper prepared for the US Agency for International Development, Battelle Memorial Institute, Pacific Northwest Laboratories.

Konstantinov, M. J. (1990). Problems in low-rank coal utilization in Bulgaria. In *Energy resources and European market economy* (ed. W. Pillmann and S. Burgstaller), pp. 60–7. International Society for Environmental Protection, Vienna.

Lencz, I. and Balajka, J. (1990). Different ways of transition of the Czechoslovak energy system to environmentally sound strategies. In *Energy resources and European market economy* (ed. W. Pillmann and S. Burgstaller), pp. 79–97. International Society for Environmental Protection, Vienna.

Levine, M. D., Gadgil, A., Meyers, S., Sathaye, J., Stafurik, J., and Wilbanks, T. (1991). *Energy efficiency, developing nations, and Eastern Europe*, pp. 1–60. Report to the US Working Group on global energy efficiency, June 1991. International Institute for Energy Conservation, Washington, DC.

MEP (1990). *National environment policy*, November 1990. Ministry of Environmental Protection, Natural Resources and Forestry, Warsaw.

MEP (1991). *National environment policy*, May 1991. Ministry of Environmental Protection, Natural Resources and Forestry, Warsaw.

Milenkovitch, D. (1991). The politics of economic transformation. *Journal of International Affairs*, **45** (1), 151–64.

NGO (1991). Statement on extension of aluminum production in Ziar nad Hronom. In *Working together*, Proceedings of the NGO conference, Prestavlky, June 1991.

OECD/IEA (1990). *The energy situation in European economies in transition*. Ad Hoc Group on International Energy Relations, April 1990. OECD/IEA, Paris.

ÖÖI (1991). Was tun, wenn ...? Brochüre des Österreichischen Ökologie-Institut, Vienna.

Österreichische Expert Commission Bohunice (1991). *Bewertung des Sicherheits der Kernkraftswerk Jaslovska Bohunice Block VI*. Band 2, Studie ersteht der Auftrag des Bundeskanzlers, Vienna.

PlanEcon (1990). *PlanEcon long term energy outlook*.

Pillmann, W. and Burgstaller, S. (ed.) (1990). *Energy resources and European market economy*. International Society for Environmental Protection, Vienna.

Radetzki, M. (1990). Energy exports from centrally planned economies. *Energy Policy*, **18** (9), 806–8.

Rippon, S. (1990). Worldwide concern over the old VVERs. *Nuclear News*, **November**, 58–61.

Russell, J. (1990). *Environmental issues in Eastern Europe: setting an agenda*. Royal Institute of International Affairs–World Conservation Union, London.

Schreiber, H. and Weissenburge, U. (1991). European plan of action on cooperation for an environmental reconstruction in Central and Eastern Europe. In *Environmental action plan for Europe*, pp. 10–13. Deutsches Institut für Wirtschaftsforschung und Institut für Europaeische Umweltpolitik, Bonn.

Thomas, S. (1990). Comecon nuclear power plant performance. *Energy Policy*, **18** (6), 506–24.

Woodard, C. *Nuclear safety in Eastern Europe*, Chapter 3. (In press.)

World Bank (1990). *Poland—the environment*. World Bank Background Paper. World Bank, Washington, DC.

World Bank (1991). *World Bank communication on the G-24 meeting on environmental protection in Poland*, September 1991. World Bank, Washington, DC.

7

The legacy of Chernobyl: the prospects for energy in the former Soviet Union

Zhores A. Medvedev

Dr Zhores A. Medvedev is not only a distinguished biochemist but also a member of the small, brave group of Russian scientists—of whom Andrei Sakharov was the most eminent—who sacrificed their careers and risked their personal liberty by openly opposing the human rights policies of the former Soviet regime. Born in Tbilisi, twin brother of the historian Roy Medvedev, Zhores Medvedev worked as a senior scientist in the Timiryazev Academy of Agricultural Science until 1962, as Head of the Molecular Radio-Biology Laboratory in the Institute of Medical Radiology, Obninsk, until 1969, and as Senior Scientist in the Institute of Physiology and Biochemistry of Farm Animals, Borovsk, until 1972. A year later, Dr Medvedev was exiled from the Soviet Union for his persistent and increasingly public opposition to the use of psychiatric treatment as a weapon against political dissidents. He settled in London, where he took up a post with the National Institute for Medical Research. Zhores Medvedev has made good use of his freedom to record the truth. He was the first scientist to expose the nuclear catastrophe—more serious, in local terms, than Chernobyl—which had occurred in 1957 at Kyshtym, near Chelyabinsk, as the result of an accident in a nuclear weapons production plant. He has written extensively about recent Soviet history: his most recent book is 'The legacy of Chernobyl' (Blackwell, Oxford, 1990).

INTRODUCTION

Seven years have passed since the Chernobyl nuclear reactor catastrophe on April 26, 1986, but its legacy for the world's nuclear energy programmes continues to grow. Outside the former Soviet Union the continuing study of the accident and its environmental consequences is motivated mainly by the natural academic interest of radiobiologists and by technical interest in the precise cause of the accident and its economic and nuclear safety implications. In 1990–1 several international conferences were held on the topic of Chernobyl in Vienna, Paris, Luxembourg, Lancaster and other venues. This field of research is likely to remain active in future decades. In the Soviet Union, and now in the Commonwealth of Independent States (CIS) which has replaced the USSR, however, the research agenda is still dominated by the health, environmental, and agricultural problems which resulted from the Chernobyl accident.

But Chernobyl also turned into a potent political symbol in the USSR, transforming the 'green movements', initially entirely concerned with the environment, into powerful nationalist forces in some republics. Before the disaster, Soviet environmental movements were weak and divided and there was no independent anti-nuclear movement. Officially sponsored groups campaigned against nuclear weapons and nuclear war, not against nuclear power. After the accident, anti-nuclear environmental movements began to emerge, and in some republics they combined national grievances with their environmental concerns. The strongest opposition to nuclear power developed in the Ukraine and Byelorussia, the two republics which had suffered most from the accident. The Ukraine also had more working and planned nuclear power stations than any other republic.

The planning, design, and implementation of all nuclear projects in the USSR was carried out in Moscow and there was little consultation with local communities about their need for power or the location of stations. When contested elections were introduced in 1989 and 1990, environmental movements organized political campaigns and elected their own representatives to the central and republican parliaments. In Lithuania, for example, anti-nuclear groups campaigned successfully against the construction of the third and fourth units at Ignalina. The first Green Party was officially registered in the Ukraine in 1990. Later that year the Ukrainian Supreme Soviet declared the republic an 'ecological disaster area' and demanded the closure of the three remaining working RBMK reactors at Chernobyl.

By April 1991 many Ukrainian and Byelorussian officials were using the term 'Chernobyl holocaust' in public discussions of the implications of the catastrophe for the future of their countries. In the campaign that preceded the referendum on independence in the autumn of 1991, Ukrainian nationalism was fuelled by comparing the famine which had resulted from collectivization in 1933, in which nearly 6 million people died in the Ukraine, with Moscow's 'radiation genocide' policy which was held responsible for the fact that so many nuclear power stations were built on Ukrainian soil. It is clear, therefore, that as far as the leaders of these two states were concerned, the accident contributed significantly to the final collapse of the USSR.

THE CONTINUING CONTROVERSY ABOUT THE ENVIRONMENTAL AND HEALTH IMPACT OF CHERNOBYL

It is not surprising that health problems tended to dominate public opinion about the Chernobyl accident and the future of nuclear power in the Soviet Union. The death soon after the accident of nearly thirty people who had suffered heavy doses of radiation, the radiation sickness of several thousand others (mostly operators, firemen, policemen, soldiers, and people who had taken part in implementing the emergency measures), the evacuation of 130 000 people from the 30 km exclusion zone around the station, and the creation of a medical radiological register of 600 000 people which included the 'liquidators' (the people who had built the sarcophagus and decontaminated the exclusion zone), evacuees, and people who continued to live in contaminated districts, caused considerable anxiety and tension in the Ukraine, Byelorussia, and parts of the Russian Federation, particularly in regions adjacent to the exclusion zone. But it was hoped that their anxiety would subside with time, as their exposure diminished.

The maximum exposure of the local population occurred during the first few weeks after the accident, when the short-lived radionuclides, particularly iodine-131, represented the main radiological threat. By 1987 exposure was already reduced and it was caused mainly by the contamination of the environment with long-lived caesium-137 and, to a lesser extent, strontium-90. Exposure continued to decline in 1988 and 1989 with the progressive decay of the radionuclides, which were also leached from the soil, and the continuing counter-measures taken by the authorities. Nevertheless, the Ukrainian and Byelorussian governments declared a massive new evacuation from contaminated areas in 1990–5 which will involve more than 200 000 rural people and require large financial and material resources.

The central medical authorities in Moscow tried in vain to convince local medical and environmental groups that the radioactive contamination had been reduced and that evacuation and resettlement programmes were no longer necessary. But these groups no longer believed anything the Central Government told them and the republics most affected formally requested international agencies to make an independent assessment of the environment and to advise them about the steps they should take to safeguard public health. The International Atomic Energy Agency (IAEA) responded by launching the International Chernobyl Project in February 1990 with the participation of the Commission of the European Communities, the World Health Organization, the World Meteorological Association, and the United Nations Food and Agriculture Organization and Scientific Committee on the Effects of Atomic Radiation. During 1990 and 1991 approximately 150 scientists and experts from well-known Western universities and institutes carried out extensive research in several contaminated districts of the Ukraine, Byelorussia, and the Russian Federation as well as in the clean districts they had selected as controls. In order to convince local people of the objectivity of the project, it was initially carried out under the chairmanship of Dr Itsuzo Shigematsu, Director of the Radiation Research Foundation in Hiroshima, Japan. When the results were released in Vienna in May 1991 (International

Advisory Committee, IAEA 1991), however, they were rejected by both official and public environmental groups in the Ukraine and Byelorussia.

By then Chernobyl had become not so much an environmental or medical problem as a political and economic issue. The responsibility of the Central Government in Moscow for the whole nuclear energy programme and for the accident was obvious. Since May 1986 the Central Government had taken responsibility for supplying the most heavily contaminated districts with clean food and paying financial compensation for the potential health hazards. Initially this applied to people, all of whom had been included in the medical register, living in random spots where the level of caesium-137 was above 15 curies per square kilometre (Ci/km^2) (higher than 555 kilobecquerels per square metre (kBq/m^2)), classified as areas of 'strict control'. Local politicians, environmental groups, and the people themselves refused to accept that the radiation danger in these spots might be diminishing. Moreover, they were campaigning to have the delivery of clean food and financial compensation extended to people living in all areas where the level of caesium was above 1 Ci/km^2 (37 kBq/m^2), arguing that low doses of radiation were dangerous. This would have meant extending the special regime to an area of more than 100 000 km^2 with a population of about 3 million people at a cost of 10 billion roubles per year. This was far more than the Government could afford.

Large areas of Sweden, Finland, Austria, and South Germany had been contaminated with caesium-137 in 1986 to levels between 37 kBq/m^2 and 150 kBq/m^2 and had needed protective measures. However, as the exposure decreased with time, the protective measures were reduced and in 1988 they ceased completely. In these countries, therefore, the cost of Chernobyl had been declining. This was the natural pattern. In the Soviet Union, on the other hand, protective measures increased over time. According to the findings of the International Chernobyl Project, 'the surveyed contaminated settlements were lower than the officially reported dose estimates' and no health disorders could be found 'that could be attributed directly to radiation exposure'

(International Advisory Committee, IAEA 1991, p. 8). There were, however, 'substantial negative psychological consequences in terms of anxiety and stress' which extended well beyond the contaminated areas. The investigators did not find a marked increase in the incidence of cancers and leukaemia, although they did not exclude the possibility of such an increase, particularly of thyroid cancer, in the future. They did not recommend the relocation of people and advised that 'food restrictions should have been less extensive.' By emphasizing the diminishing threat of environmental radiation exposure and recommending the gradual reduction and eventual abolition of protective measures, the findings confirmed the views of the Central Government and contradicted those of local officials. As a result, local officials rejected the findings.

By 1991 differences about the Chernobyl problem had also begun to be voiced in Moscow. The Ministry of Health insisted that the radiological threat was diminishing, whereas the Chernobyl Committee created by the USSR Supreme Soviet took the opposite view. In fact, the curious thing about the treatment of the consequences of Chernobyl was that the areas considered to be contaminated (and therefore the number of people classified as victims) kept growing. Yu. A. Israel, the chairman of the State Committee on Hydrometeorology, for example, reported in a review article in *Pravda* on 26 April 1991 that the radiation monitoring services had identified many new spots contaminated with caesium-137 in the Ukraine, Byelorussia, the Russian Federation, and even in Georgia in 1990. Large areas contaminated with caesium-137 levels between 1 and 5 Ci/km² had been found in Kaluga, Orel, Ryazan, Smolensk, Tula, Tambov, Leningrad, Voronezh, Krasnodar, and other regions along the western coast of the Black Sea. The total area with this level of contamination was reported to be 103 200 km². The area contaminated with levels of 5 to 15 Ci/km² of caesium-137 was enlarged from 1409 km² in 1989 to 17 610 km² in 1990.

In 1989 the State Committee on Hydrometeorology had published maps to show the areas contaminated with caesium-137 above 15 Ci/km². In 1990 new maps were published which also identified areas contaminated with levels between 5 and 15 Ci/km^2. By 1991 the maps included areas in which the level had been up to 5 Ci/km^2 in 1990 because the people who lived in those areas were now officially considered victims of Chernobyl and eligible for social and financial compensation.

The USSR Supreme Soviet passed a 'Law on the social protection of citizens who suffered as a result of the Chernobyl catastrophe' on 12 May 1991 which introduced a new classification of contaminated areas. It specified that only those areas where the additional Chernobyl-related annual dose of radiation did not exceed 1 millisievert (0.1 rem) were considered safe and therefore did not require any protective counter-measures. Anyone exposed to higher annual doses was given the status 'victim of the Chernobyl catastrophe'. This status was also given to evacuees, people who had taken part in the decontamination work in 1986 and 1987, and personnel of the Chernobyl nuclear power station. All these people were eligible to some or all of a list of benefits: financial compensation, free medicine and dental treatment, free annual treatment at sanatoria, extended holidays, free transport, priority for housing and education, early retirement on full pension with a 30 per cent supplement, income tax relief, and interest-free credit. The law gave 29 different benefits to the victims of radiation syndrome, 17 to people who had worked on the clean-up operations and evacuees, 15 to people who lived in areas contaminated above 5 Ci/km^2, and 7 to the largest group— the 4 million people who lived in areas with contamination levels of 1 to 5 Ci/km^2. In addition to 30 roubles per month compensation per person, these people were to receive a 15 per cent supplement to their basic salaries, extended paid holidays and maternity leave, regular medical checks, and priority in higher education. All cases of leukaemia, thyroid cancer, and other malignancies in children in areas with a caesium-137 level of over 5 Ci/km^2 were to be treated as radiation linked and compensated in the same way as those who suffered acute radiation syndrome.

The law was also extended to victims of previous radioactivity accidents in civilian and military nuclear facilities and nuclear weapons tests. This meant that people who had been evacuated from areas contaminated in 1957 as a result of the Kyshtym accident in the Urals and many settlements around the nuclear weapons test grounds in the Semipalatinsk region of Kazakhstan would now receive the same treatment as the victims of Chernobyl. The implication of this legislation was that the Central Government undertook to spend about 20 billion roubles on compensation and benefits—or the same amount of money as was allocated in the budget each year to the entire Soviet health service. The intention was clearly to make concessions to local demands and to subsidize the Ukrainian and Byelorussian social and health programmes, but the amount of money involved was unrealistic, particularly since an increase in annual dose of one millisievert could hardly be considered as creating a measurable new risk. When the Central Government collapsed in December 1991 and the USSR disintegrated, the law became irrelevant.

THE ECONOMIC COST OF CHERNOBYL

Chernobyl became an economic calamity for the whole of the Soviet Union. In the rest of Europe the economic cost of Chernobyl (in other words, the cost of radiation protection and control measures and of compensation to farmers for lost agricultural products) was highest in 1986 (about US$ 1 billion). After this the cost declined steadily. In the USSR, on the other hand, the cost increased annually from 2 billion roubles in 1986, to 8 billion roubles in 1988 and 17 billion roubles in 1991, when the USSR Supreme Soviet passed the new decree. The declaration signed in Minsk on 8 December 1991 which replaced the Soviet Union with the Commonwealth of Independent States included a clause stating that financial support for the Chernobyl programme would remain a priority.

Independent attempts to assess the real costs of Chernobyl up to the year 2000 have suggested figures ranging from 170 to 215 billion roubles (in 1986 prices). But these calculations did not include the remaining cost of decontaminating the exclusion zone. When I visited the site in 1990 and 1991 I found that there were nearly 800 very primitive temporary nuclear cemeteries or disposal sites around the Chernobyl plant containing more than 5 million curies of long-lived radionuclides. Radioactivity from these sites continues to contaminate the groundwater. The nuclear reactor, now covered by a concrete sarcophagus, still contains 650 kg of plutonium, 180 tonnes of uranium, 7 million curies of radiocaesium, 6.1 million curies of radiostrontium (strontium-90), and many other radionuclides in a very unstable condition. A total of nearly 50 million curies of radioactivity remains dispersed under the rubble. Moreover, the sarcophagus which covers the reactor is not expected to last for more than twenty years. A new protective structure which is larger and heavier will have to be built to encapsulate it, or else the sarcophagus will have to be removed so that the destroyed reactor can be dismantled bit by bit by special robots and buried properly. Nobody knows how to do this, how long it will take, or how much it will cost. The Ukrainian parliament recently appealed to the United Nations to create a special task force to dismantle the sarcophagus and the remaining three RBMK-1000 reactors on the site since the new Ukrainian Government has neither the money nor the knowledge to do the job.

THE SOVIET NUCLEAR ENERGY PROGRAMME

After extensive consultations with scientists and economists, Gorbachev presented an All-Union Energy Programme in October 1985 in which nuclear energy was given priority. In 1985 the total production of electricity in the Soviet Union was 1544 billion kilowatt-hours (kWh). Only 10.8 per cent of this output came from nuclear power stations (*Narkhoz* 1985), far less than the proportion of nuclear-generated electricity in the United States, the United Kingdom, West Germany, France, Spain, or Sweden and less than the proportion of nuclear power in the national electricity production of several East European countries. The new programme envisaged an increase in nuclear-generated electricity to 22 per cent by 1990, and 40 per cent by the year 2000. These figures were incorporated in the 1985–90 Five Year Plan which was approved by the 27th Congress of the Communist Party of the Soviet Union in February 1986. This was an attempt to keep up the pace of constructing nuclear power stations which had been reached in the previous five years, when the total power of nuclear stations was increased from 12.5 million kW in 1980 to 28.1 million kW in 1985 and the annual production of nuclear-generated electricity rose from 72.9 billion kWh to 167 billion kWh. The Chernobyl accident put paid to this plan, however, particularly because nearly half the new nuclear reactors were planned to be the same RBMK-1000 type as those at Chernobyl. In 1987 all construction on reactors of this type was stopped.

Despite Chernobyl, there was a powerful economic incentive behind the nuclear programme. The energy crisis which developed in the USSR in 1987 slowed down economic development (the growth of gross national product (GNP) was only 1 per cent in 1987) and made it necessary to increase the consumption of oil and gas. This was a difficult option because of transportation problems. Most of the coal deposits were in Siberia and Northern Kazakhstan, and most of the oil production was in Western Siberia. The single rail-link between the European part of the USSR and Siberia could not cope with the increased transportation of coal and the Siberian–European oil pipeline had limited capacity. Thus when a revised Energy Programme was adopted in October 1988, it retained a less prominent nuclear option, despite strong pressure from the very vocal anti-nuclear movement and general popular anxiety about nuclear power stations.

In 1988 all Soviet electric power stations with a total power of 338 million kW produced 1705 billion kWh of electricity, 216 billion kWh of which was generated by nuclear power stations (*Narkhoz* 1988). The total power of nuclear energy was 35.4 million kW in 1988. If the new energy programme were fulfilled, the total power of electricity generators would increase by 200 million kW, with nuclear power stations responsible for

60 million kW (fossil-fuel and hydroelectric stations were planned to contribute 70 million kW each). This would mean tripling the number of nuclear reactors in the Soviet Union within twelve years. Like other elements of the programme, this was clearly unrealistic. In fact, the rate of increase of production of oil, coal, and natural gas slowed down in 1985–8 and there were strong objections to new hydroelectric projects because they required the resettlement of large groups of people.

There was powerful opposition from environmental and nationalist movements to the new programme. In December 1988 the Armenian earthquake forced the Government to shut down the Armenian nuclear power station: although the two 500 megawatt (MW) reactors of the Armenian Atomic Energy Station (AES) had not been damaged, if the epicentre had been closer, a new 'Armenian Chernobyl' would have occurred very close to Yerevan. The Armenian Government demanded that the AES should remain permanently closed. As a result, the energy programme was badly damaged and anti-nuclear sentiments gained further momentum. Soon the Kuban and Crimean nuclear power stations which had been under construction for several years were also closed because their sites were considered seismically unstable.

There were about twenty-five nuclear reactors under construction in the Soviet Union in 1989 and each project was attacked. The anti-nuclear lobby gained strength from the democratization of the country and the decentralization of administrative power which followed the election of the first Congress of People's Deputies of the USSR and the Supreme Soviet. Environmental issues and nuclear energy projects became emotive election issues in many areas and many deputies were elected for their anti-nuclear views. This was particularly the case in the Ukraine, Byelorussia, and the Bryansk and Chelyabinsk regions of Russia and Kazakhstan, where nuclear power was associated either with nuclear disasters or with weapons tests.

In 1990, when a more democratic electoral system was introduced for republican, regional, and district elections, the environmental and anti-nuclear campaigns gained further influence.

Opposition to the Central Government and to the Communist Party of the Soviet Union was often united around environmental issues. Local public 'hearings' were organized in many regions over controversial construction projects, some of them already completed. Newly elected parliaments passed anti-nuclear resolutions and redirected funds allocated earlier for nuclear power stations.

The anti-nuclear campaign was well organized and coordinated. It was also surprisingly effective. Official figures published by the USSR Ministry of Atomic Energy in March 1991 indicated that between 1987 and 1991, sixty nuclear reactors were either shut down permanently or halted in the construction stage. The Rostov, Crimean, Tartar, and Bashkir AES were frozen when they were nearly completed; the South Ural AES with three fast reactors was halted after a regional referendum at an early stage of construction; new reactors for Smolensk, Kalinin, Zaporozhye, Khmelnitsky AES were cancelled despite large investments already made to build them. Almost all the thermal nuclear stations which were intended to heat large cities were also frozen (the Gorky, Voronezh, and Odessa stations had already been completed and tested). Many projects were abandoned at the design and planning stage. This meant the loss of 100 million kW in the immediate future (*Izvestiya*, 23 March 1991), which is equal to the power of one hundred reactors of the standard 1000 MW type (this was equal to the total power of all Soviet electric power stations in 1964). Huge sums of money and significant amounts of resources and high technology equipment were wasted. The situation was further complicated by the fact that oil and coal output which had peaked in 1988 began to decline in 1989, and was seriously reduced in 1990 and 1991.

The result of this trend was not surprising. The energy crisis which had begun to be felt in 1987 developed into such an acute shortage of electricity that it became necessary to limit the sales of electricity to all consumers, whether industrial, private, or for export (for example to Bulgaria, Hungary, and Poland). The nuclear energy debates were resumed and the nuclear industry and a rather influential network of nuclear energy research institutes were mobilized to produce the necessary arguments to change public opinion. The energy crisis was particularly severe in Armenia, Georgia, and some regions of the North Caucasus where many industrial plants had to be closed to save electricity. There were also serious shortages of electricity in Siberia and the Far East. The price of electricity for private and industrial consumers was raised. These economic factors finally began to change public attitudes towards nuclear energy. But bad luck struck the nuclear industry again at the Chernobyl AES on 11 October 1991.

In 1986 the Soviet and Ukrainian Governments had mobilized nearly half a million workers and soldiers to build the protective cover for the destroyed reactor in Unit 4 of the Chernobyl AES and to decontaminate the site around the power station so that the three undamaged RBMK-1000 reactors could be put back into operation. The electricity produced by the Chernobyl AES was essential to supply the population of Kiev and local industry. Although this emergency programme was successful, the costs, particularly in terms of the health of the people who carried it out (now known as 'liquidators') who were exposed to very high doses of radiation, were extremely high. By November 1986 the cover, a huge concrete sarcophagus which encapsulated reactor 4, had been completed and reactors 1, 2, and 3 resumed operation. Three million kilowatts of electricity were available for Kiev just in time for the winter.

On 11 October 1991 an accident in the machine hall of the Chernobyl AES caused a fire of hydrogen and oil. The fire was so powerful that the metal structure of the machine hall began to melt and the roof collapsed. The flames were so intense that four times as many fire brigades were required to extinguish them than had been used during the first accident on 26 April 1986. Fortunately, in 1991 there was no escape of radioactivity and no damage to the reactors was registered. But the fire rekindled all the old fears about nuclear energy. As a result, the Ukrainian Supreme Soviet passed a resolution on 29 October 1991 to close down the entire Chernobyl AES permanently (it had already been shut down temporarily because of the damage to the machine hall). The Ukrainian Parliament proposed to dismantle the entire site, together with the sarcophagus, and to dispose of all the nuclear materials. Since no one in Ukraine or Russia possessed the knowledge and technology for such an enormous task, the Ukrainian Parliament appealed to the United Nations for funds and technical assistance. The rather naïve expectation was that the United Nations would be able to create international technical teams which, like UN peace-keeping forces, could be sent to different parts of the world to direct and subsidize the decommissioning and dismantling of nuclear power stations. In January 1992 energy shortages in the Ukraine forced the Ukrainian Government to sign a contract to import oil from Iran; this was the first time in the history of the former Soviet Union that oil was to be imported rather than exported.

The winter of 1991–2 was, fortunately, very mild. However, political factors—the disintegration of the USSR and the formation of the CIS from eleven of the former Soviet republics (Estonia, Latvia, Lithuania, and Georgia all opted for full independence)—made the situation very complex. Traditional economic links between the regions were severely affected and the replacement of the single system of distribution of resources, commodities, produce, and energy by a new commercial system of trade put many areas, which had previously received electricity from the single consolidated energy grid, in jeopardy. Many Estonian factories and plants, for example, had to be closed during the winter because of energy shortages. Problems with central heating made it necessary to move many thousands of people from towns to villages, where firewood could be used to heat houses. In Armenia, where the shortages were exacerbated by conflict with Azerbaijan, almost all industrial enterprises had to be closed down in January 1992. Armenian villages were unable to cope with the lack of heating and there were reports that in some villages fruit trees had been cut down for use as firewood. In many towns in Western Siberia the use of firewood was the only way that the winter could be survived.

The energy crisis of 1992 once again changed public attitudes to nuclear power. The Armenian Government, for example, wants to reopen the nuclear power station which was closed after the earthquake but it cannot do this immediately because essential equipment has gone missing or been damaged during three years of neglect. But there is a further problem—when the Soviet Union disintegrated, the central Ministry of Nuclear Power disappeared with it. Now each republic has to develop its own energy programme and if it opts for nuclear power, it will have to negotiate with the Russian Federation, since Russia is the only member of the CIS with the competence to build and use nuclear power stations. Energy supplies may well turn out to be the main factor in the economic cooperation between the former republics of the USSR.

A BRIEF REVIEW OF THE ENERGY SITUATION IN THE FORMER SOVIET REPUBLICS

In 1991, when Gorbachev's second energy programme was undermined, his Government introduced a third programme, based this time on an attempt to increase the production of coal, oil, natural gas, and the construction of new hydroelectric stations. Gorbachev seemed to realize that although many conflicts threatened the continued existence of a unitary state or a new federation, the joint energy system which had been created over many decades could potentially act as the main unifying force, keeping the republics together. In fact, he was unrealistic and it was far too late. The production of oil in the USSR was declining for the third year running, after the peak year of 1987 (624 million tonnes). Coal miners' strikes, which had adversely affected coal production in 1989, reappeared in 1991. During the winter of 1990–1 there were increasing shortages of fuel and electric energy. The situation varied from republic to republic, and this added to the tension between the Central Government and republican authorities.

Four of the republics of the new Commonwealth produce energy surpluses which they can export: the Russian Federation produces a surplus of oil, gas, and coal; Kazakhstan has a surplus of coal and oil; Azerbaijan has a surplus of oil and gas; and Turkmenistan has a surplus of natural gas. The other republics have to import energy. In 1990 Armenia could provide only 4.3 per cent of its energy requirements from its own resources; Byelorussia produced 11.5 per cent of its needs; Georgia produced 19.5 per cent; Latvia produced 8.1 per cent; Moldova produced 1.5 per cent; and Lithuania produced 24 per cent (Khrilev 1991). Even Ukraine, the most industrially developed republic after Russia, was self-sufficient for only 41 per cent of its requirements. In order to import energy at world prices, Ukraine would need about US$ 10 billion per annum, while Byelorussia would need US$ 4 billion. Neither Government can afford this.

In the former USSR energy resources were shared regionally. Turkmenian natural gas, for example, was distributed to the other republics of Central Asia and it was also pumped to the All-Union gas pipeline system to be delivered to the Ukraine. Oil- and gas-rich Azerbaijan was linked by oil and gas pipelines to Armenia and Georgia and provided these republics with petrol, diesel fuel, and heavy residual fuel for power stations. The disappearance of the centrally planned economy and the replacement of the administrative-command system of distribution by trade based on market principles has put an end to the regional sharing of energy resources. Turkmenistan now plans to sell a significant proportion of its natural gas to Pakistan, Iran, and Turkey which will pay for it in dollars. Azerbaijan has stopped pumping gas and oil to Armenia because of the conflict over Nagornyi Karabakh. Georgia has not joined the CIS and faces an energy crisis.

Uzbekistan, the most populated republic in Central Asia (more than 20 million inhabitants), has an acute shortage of fuel and electricity. Its electricity stations produce only 11 million kW, about 3 per cent of the total produced in the CIS, and it has no energy reserves, apart from solar and wind energy. Its hydro-electric power stations (the largest is Charvak) have a dual role, since their dams are also used for irrigation. This means that they are not used properly to produce electricity in the growing season. Moreover, these dams have created one of the most notorious environmental problems in the world—the Aral Sea disaster. Any attempt to save the Aral Sea involves increasing its water input and therefore reducing irrigation. Water is the most valuable commodity in the area and this makes nuclear-generated energy an unlikely option for Central Asia: nuclear power stations require at least three times more water than power stations that use fossil fuels.

The republics which suffered most from the Chernobyl accident—Ukraine and Byelorussia—and which were the first to stop the construction of nuclear power stations, are also expected to suffer acute shortages of electricity for domestic and industrial use unless the anti-nuclear trend is reversed. Before the collapse of the USSR, they were expected to increase their consumption of coal from Western Siberia (Kuzbas) and North Kazakhstan (Karaganda). Now that the payments for this coal are to be based on market prices, it is likely to be too expensive. The projected special 'coal rail line' between the eastern and western parts of the USSR has been shelved because it is too expensive. For Byelorussia, importing coal from Poland will cost less than buying it from Siberia.

In 1991 there were fifteen nuclear energy reactors under construction in the USSR and several had already been completed (for example Rostov and Gorky). It is very likely that they will all be completed and put into operation. The Chernobyl syndrome will begin to wane when shortages of food and energy become acute. In the five years since the Chernobyl accident, the nuclear industry has been modernized. The manufacturers of RBMK-1000 reactors and graphite-moderated military reactors switched to the construction of new pressurized water reactors similar to those which have become popular in the United States and Japan. The new generation of VVER-1000 reactors will be exported from Russia to Eastern Europe, Finland, North Korea, Iran, Bangladesh, Syria, India, and Pakistan. Russia also plans to increase its export of enriched uranium and plutonium for use as nuclear fuel. Moreover, as a result of the conversion of the military industry and the secret nuclear facilities in Kyshtym and

Krasnoyarsk, Russia has recently offered to reprocess spent nuclear fuels and to store nuclear waste from other countries. This is intended as a commercial venture which will be important for the future of the nuclear industry.

OUTLOOK

Chernobyl was a crushing blow to the Soviet nuclear programme and for Soviet nuclear science. However, given the demands of economic development and the insuperable difficulties of finding alternative sources of power, the restoration of the nuclear energy option is inevitable in Russia and other republics. The industrial development of Siberia, Kamchatka, Chukotka, and the Far East of Russia will be extremely difficult without nuclear power stations. The millions of square kilometres of permafrost in these regions make it impossible to consider constructing roads to transport fossil fuels. The many thousand redundant scientists, engineers, and technicians from the military nuclear industry could be employed on civilian nuclear energy projects. The nuclear energy industry brings Russia important export revenues. Moreover, Russia could export electricity as well as nuclear technology. In the former Soviet Union the export of electricity increased from 20 billion kWh in 1980 to 40 billion kWh in 1989. This represented only about 2.3 per cent of the total production of electricity in the USSR (Ratnikov and Glukhovsky 1991) and it was exported to Hungary, Bulgaria, Poland, Austria, Finland, Turkey, and Mongolia. In future, other members of the CIS will join the list of countries that import electricity from Russia. In 1991 the export of electricity dropped to 21.5 billion kWh. This means that the level will be low in 1992, despite the desperate need for foreign currency (50 billion kWh is worth about US$ 3 billion).

Russia is second only to the United States in its experience of the problems of nuclear energy. Neither nuclear superpower can afford to develop its energy systems without using nuclear power. Although the Chernobyl accident and its aftermath provoked an energy crisis that contributed to the collapse of the Soviet Empire, the development of the new Commonwealth will be impossible without a significant increase in the production of energy.

REFERENCES

International Advisory Committee, IAEA (1991). *The international Chernobyl project*, an overview, conclusions and recommendations, surface contamination maps, assessment of radiological consequences and evaluation of protective measures. IAEA, Vienna.

Israel, Yu. A. (1991). Panorama osoboi zony. *Pravda*, 26 April.

Khrilev, L. S. (1991). Chto zhdet respubliki. *Energiya*, **7**, 44–7.

Narkhoz (1985). *Narodnoye Khozyastvo SSSR v 1985 godu*. Finansy i Statistika, Moscow.

Narkhoz (1988). *Narodnoye Khozyastvo SSSR v 1988 godu*. Finansy i Statistika, Moscow.

Ratnikov, V. and Glukhovsky, M. (1991). Kilovatt-chas na eksport? *Energiya*, **6**, 6–8.

8

Nuclear power and the environment

Walter Marshall

When the history of nuclear power in the United Kingdom is written, the name of Lord Marshall of Goring, FRS, will have more index references than most. He has been involved with the industry, both as a distinguished scientist and as an energetic and sometimes controversial manager, for most of his career. He joined the Atomic Energy Research Establishment (AERE) at Harwell as a research physicist in 1954 and remained there, apart from two years of research in the United States, until 1975; by then he had been Director of the AERE for nine years, following six years as Head of the Establishment's Theoretical Physics Division. During his term as Director at Harwell, Walter Marshall served simultaneously as Chief Scientist at the Department of Energy (1974–7) and as Director of the Research Group of the United Kingdom Atomic Energy Authority (UKAEA) (1969–75). He was appointed Chairman of the UKAEA in 1981 and then moved a year later to take over the Chairmanship of the Central Electricity Generating Board (CEGB). He occupied that position until 1989 and was consequently in the thick of the argument, during Mrs Thatcher's premiership, over the proposed privatization—and consequently fragmentation—of the British power generating industry. Lord Marshall (as he had become in 1985) opposed privatization with characteristic vigour, on economic and managerial rather than political grounds: he lost the battle, but many of his arguments were vindicated when the Government was obliged to withdraw the nuclear power industry from the privatization programme. Shortly after the Chernobyl disaster and partly because of it, Lord Marshall founded and stimulated wide international support for the World Association of Nuclear Operators, of which he is now Chairman. The Association's objective is to raise standards of the training of, and encourage the exchange of information between, those responsible for operating the nuclear power industry worldwide.

Recently, in the *Financial Times*, a share analyst commented that BP had an expensive and adventurous policy looking for new sources of oil. The analyst went on to discuss what that meant for BP's share prices, but the conclusion I drew was that the supply of oil was not as comfortable as BP would like it to be.

Indeed, most discussions about the supply of fossil fuels emphasize the high reserves in the Middle East and the political uncertainty of retrieving them, and the high reserves in Siberia and the difficulties, both technical and political, of developing them.

I am uncertain of the significance or value of these particular stories, but they reinforce the view which I have held for a long time, and which was at one time popular, that accessible sources of convenient fossil fuels, like oil and gas, will become rarer some-time in the first half of the next century. I therefore remain convinced that nuclear power is the ultimate source of energy for mankind. Its use will start to grow in the early decades of the next century and will dominate by the year 2050. It is therefore possible that our children, likely that our grandchildren, and probable that our great-grandchildren will rely upon nuclear power as their major source of energy.

That being the case, let us try and understand the environmental impact of nuclear power and let us try and do that now so that we can all get used to the idea before we are obliged to accept it.

Nuclear power is not popular nowadays and that unpopularity is due almost entirely to misunderstanding and to the fear that the general public has of all things nuclear. For the general public, their introduction to nuclear science was the explosion of atomic bombs at the end of the Second World War. That was a frighten-ing beginning. Furthermore, all people know that the production of nuclear energy inevitably involves the production of radio-activity and radiation, and people fear radioactivity and radiation because they know they cause cancer. Of all diseases, cancer is the one that people fear the most. In my view all these fears are irrational and emotional but nevertheless they exist and, because they exist, the industry strives for a safer and safer technology which in practice means a more expensive technology.

In my opinion, nuclear power will not make much new progress until the benefits and the risks of the technology are seen in a fair perspective, and therefore I shall concentrate on this aspect of the subject. I shall not bore you with details of nuclear engineering but instead I will describe, from a fundamental point of view, why nuclear power must be the most environmentally friendly source of energy.

I shall discuss the environmental impact of nuclear power by using three different approaches: comparative, philosophical, and mathematical.

Let us begin with a comparative approach.

How can we determine whether a technology is environment-ally friendly or not? I suggest we do this by looking at the size of man-made pollution against the natural production of similar materials in nature. As a rough empirical rule, if what we produce is a small fraction of what God produces already, then we ought to be relaxed. But, if mankind's production is comparable with that which God produces then, a priori, we have reason to fear we might be upsetting the balance of nature. Let us examine four environmental problems this way: the production of acid rain worldwide, the ozone hole over the polar regions, the developing greenhouse effect, and, finally, radioactivity produced by nuclear power. Let us get a relative measure of pollution in these areas produced by man and by God.

Acid rain is produced primarily through the release of sulphur dioxide from the combustion of coal and oil. Acid rain is therefore a pollutant produced worldwide by coal-fired power stations, oil-fired power stations, motor cars, and most forms of transport. Sulphur dioxide is also emitted naturally from volcanoes and from the oceans and, in the Northern Hemisphere, it has been calculated that mankind produces 10 units for every 1 unit produced by God. This ratio of 10 to 1 alerts us immediately to a potential environmental problem. We are clearly disturbing the balance of nature and we need to make a careful judgement whether that balance is harmful or not.

In this case, society has determined that the production of acid rain is harmful. As a result, throughout Europe and America, coal and oil-fired power stations are being equipped with desulphurization equipment and the burning characteristics of those power stations are also being modified to produce less nitrogen oxides. It is more difficult to eliminate sulphur from the combustion of fossil fuels in motor cars and household boilers and in industry generally, and therefore those problems have been set aside.

There is a general lesson we can learn from this acid rain story. We first have to determine whether we are disturbing the balance of nature and I am satisfied that the answer to that is yes. Next we have to ask the more difficult question: 'Is this disturbance causing environmental harm?' It is certainly causing environmental harm by acidifying lakes and rivers and producing a bad environment for fish. But most people think that acid rain affects trees although, in my opinion, that assertion remains unproven. Of course acid rain does affect trees in localized areas of Poland and Czechoslovakia but it does seem likely that the tree damage that is found throughout Germany is primarily a result of magnesium deficiency. The lesson we should learn from this is that we should examine all steps of our argument properly. This particular nuance of that argument will become important as oil and gas grow more expensive in the future and once again it becomes fashionable to think of burning coal. In our reference comparison we can therefore mark acid rain pollution relative to 'God's pollution' as 10:1.

The ozone hole is produced by mankind's use of CFCs (chlorofluorocarbons). These materials are not produced naturally at all and therefore the ratio of the pollution by man and God is infinite. Once again, we have disturbed the balance of nature and we must not be surprised if that produces unfortunate effects.

In this case, it is now established that the ozone hole permits ultraviolet radiation through to the Earth's surface which can cause skin cancer. Therefore there is a worldwide determination to reverse this man-made pollution. Whether that can be done or not depends entirely upon the commitment of large developing countries like China and India. They need to avoid the use of CFCs and use hydrofluorocarbons (HFCs) instead. But doing this for all refrigeration equipment is a difficult task.

I do not know whether this worldwide programme will be successful or unsuccessful but, in this case, I think the politicians have got the science and the political reaction right. Let us hope they remain committed to it. Our reference figure for the CFCs is therefore infinity to one.

Let us now turn to the greenhouse effect. The burning of fossil fuel releases into the atmosphere carbon dioxide and this extra carbon dioxide will cause a small average increase in temperature of the Earth's surface. The increase in carbon dioxide concentration is very well documented for the last thirty years. It has increased by 10 per cent and therefore, over the century as a whole, we might guess at an increase of, say, 25 per cent of the natural amount of carbon dioxide in the atmosphere.

In this case the disturbance to the balance of nature is less certain and the consequences of this disturbance are not clear. If you are an optimist, you will decide that you should be cautious about the greenhouse effect; if you are a pessimist you will decide to be alarmed. Depending upon which judgement you make, you will either tolerate or oppose the burning of fossil fuels to make electricity.

Now let us use the same comparison methods to look at radio-activity. There are three important man-made sources of radio-activity: one is in medicine and that seems to worry no one, another is from nuclear accidents, and the third is from nuclear waste which is the end product of making nuclear electricity. Strangely enough, it is nuclear waste that worries people the most and therefore I will concentrate on that.

Let us suppose that the world had a large nuclear power programme producing a large amount of man-made nuclear waste and let us compare that to the nuclear material and the radiation which exist in the world anyway. We will discover that the nuclear waste produced by an enormous nuclear power programme is absolutely trivial compared with the natural radio-activity left in the Earth by God.

Let us suppose our worldwide nuclear power programme consists of 250 reactors, each of 1000 megawatts, running for forty years. As the spent nuclear fuel is taken out of these reactors, let us also imagine that it has cooled for ten years and we then assess the total amount of radioactivity produced. We will discover that we have produced 70 000 million curies of radioactivity. For the purpose of this paper we do not need to understand what a curie means, we only need to understand that it is a measure of radioactivity and is therefore a suitable measure of the man-made pollution produced by this large worldwide nuclear programme. Let us now compare that to natural radioactivity.

We know the Earth was formed approximately 4500 million years ago and it was then a very radioactive place. Most of that radioactivity has decayed but a very significant amount remains today, mainly as uranium, thorium, and potassium.

These radioactive materials generate radiation and we know from fundamental principles that the radioactive atom itself cannot cause us harm except by emitting radiation and that the properties of that radiation must be independent of its source. Therefore we can make direct comparisons of man-made or natural radioactivity.

A standard textbook on geology will tell you that most of this natural radioactivity in the Earth is concentrated near the surface—in the continental crust rather than in the Earth's mantle or core. The radioactivity is therefore not deep in the Earth or underneath the oceans, it is right here in the Earth where we live and virtually everything we look at or touch is radioactive. The concrete in any building is radioactive, the steel framework of the building is radioactive, the streets are radioactive, and all that radioactive material emits radiation which affects us. We can calculate the total radioactivity in the Earth's crust to produce 500 million million curies of radiation.

So, in this case, the ratio of man-made pollution to God's natural pollution is a ratio of 1 to 7000. This ratio is very comforting. It tells us that we are not disturbing the balance of nature appreciably. Theoretically we could afford to distribute this nuclear waste evenly throughout the world and it would be a small pollution effect if assessed in a detached, factual way. But that, of course, would be irresponsible because we can do better. In practice we can confine the nuclear waste and prevent it from affecting mankind at all. Therefore we concentrate it, surround it with concrete, and bury it deep underground. By these means we can make the risks to mankind smaller by at least a factor of 1000 so the overall comparison of risk compared with natural radio-activity is even more favourable to nuclear power.

Furthermore, a straightforward comparison of radioactivity, i.e. curies, overstates the risk from man-made radiation. This is because the measurement of radiation in curies only estimates the potential risk from radioactivity. The actual risk depends on the ease of the pathway of the radioactive material to the human body.

By a remarkable coincidence, this pathway is relatively easy for God's natural radiation because radon is a gas and can therefore easily get into our lungs, and radium is soluble in water and can easily get into our food supply. In contrast, most of the fission products produced in a nuclear reactor do not have those un-fortunate chemical characteristics. Exceptions to this are iodine and caesium but in practice all the other man-made radioactivity hardly produces any risk at all.

For all these risks, the risk to mankind from our big nuclear power programme as compared with natural radioactivity are probably a million times less than the comparison I made earlier in this paper. On the other hand, perhaps in assessing the effects of natural radioactivity I should not have taken all the natural radioactivity in the continental crust. Maybe I should only have taken the top 100 metres. That would give us a factor of 400 going in the opposite direction. We might well estimate that a combina-tion of all these factors put together might give a safety factor of 1000 compared with the previous comparison I made, so we end up by saying that the risk to mankind of a big nuclear power pro-gramme as compared with the risk to mankind coming from God's natural pollution is a ratio of 1 to 8 million. That is even more reassuring than the previous comparison.

Given those numbers, it is hard to see why people become emotional about nuclear waste at all. Throughout the world there is a well-known NIMBY (not in my back yard) phenomenon. This has no foundation whatever as a logical argument.

We have now completed our comparative table where we have looked successively at acid rain, the ozone hole, the greenhouse effect, and nuclear waste. Without any doubt the nuclear comparison is the most reassuring. To complete our comparison we should of course look at nuclear accidents but I will look at that from a different point of view later in this paper so, to avoid repetition, I will not give a similar comparison here.

As a second contribution to this paper I promised to give you a philosophical discussion. People are frightened of radioactivity because they cannot see it, they cannot taste it, they cannot detect it in any way, and therefore they cannot avoid it. However, if you think about the matter more carefully, that is exactly why we should not worry about it. The human body cannot detect radiation because it is not necessary for the human body to detect it. It is not very important.

It was very important for primitive man to see dangerous animals approaching him and therefore God or natural evolution, you may choose your own vocabulary, gave him eyes to see with. If all primitive men had been blind, mankind would have died out. It was important that primitive man had ears to hear. If all primitive men had been deaf, mankind would not have survived. It was important that he could smell, taste, and feel and, without all these senses, mankind would not have evolved as it is. If radiation really were dangerous, then mankind would not have evolved at all. Furthermore, God has not given us an ability to detect radiation with our senses because it is not actually worth bothering about and we cannot avoid it anyway.

We actually know a great deal about the risks of radioactivity and we can therefore estimate the risks from radiation, from other forms of cancer, and from other diseases. For the UK we estimate that for every 1000 people, 1 will die from natural radiation, 249 from other causes of cancer, and 750 from something else. Therefore, if we do want to worry about dying, we ought to worry about other causes of death or other causes of cancer. We believe the latter to be some mixture of chemicals, diet, or chance but it is certainly not productive to worry about the risk of death from natural radiation.

The risks of cancer are less in the USSR or for primitive man than in the UK. That is not because a Soviet citizen or a primitive man is less prone to cancer, but because he is much more prone to die of something else.

Sometimes people become confused about the risk of man-made radiation because people have chosen to present arguments in an extreme and invalid form. There are so many examples of this that I cannot quote them all. I shall only quote the most famous one where an antinuclear campaigner argued that pluto-nium, a man-made material, was the most dangerous substance known to mankind and a plutonium ball the size of an orange would be enough to kill off all mankind. That of course is absolute rubbish. Our plutonium orange does contain enough plutonium atoms and does in fact emit enough alpha particles to kill off all mankind but we would need a 'Maxwell demon', i.e. someone who could defy the laws of thermodynamics, to get the plutonium atoms from our orange ball into the lungs of so many people. This is therefore an impossible feat.

In my opinion, the counter argument was best explained by a lady who happened to be a Catholic nun teaching in America. She actually worked in a Carmelite Order and she left the Order by a special dispensation from Pope John XXIII. When she gave the counter argument she had therefore ceased to be a nun but, nevertheless, we all expected her speech to be phrased in careful, reserved tones. To be frank, we expected it to be a little dull. Instead she produced a startling analogy. She pointed out that each and every day, a healthy young American man produced enough sperm to make pregnant every woman of child-bearing age who was alive in the American continent. That, however, did not make him the most dangerous man alive because the pathway from that young man to all those millions of women could not be imagined and the chance of that one man fathering a billion children could safely be ignored.

Here again we come to a philosophical point. When God made the Earth, he did not put the radioactivity in the centre of the Earth or under the oceans or a long way from mankind, he put it all in the continental shelf, more or less where we are. Therefore, roughly speaking, he put it in exactly the areas where we, mankind, live. Either he was extremely irresponsible or he has known all along something which mankind finds difficult to grasp: radiation does not matter very much. God could therefore afford to surround us with radioactivity and irradiate us at all times and in all places because the adverse effect upon us is extremely modest. The effect of a nuclear power programme upon us would be even less.

I promised you a third section, a mathematical argument concerning the environmental impact of nuclear power. In the arguments I have used so far, I have relied upon the consensus judgement of the risks of radiation to mankind but there are some difficulties with this.

Most of our information concerning the effects of radiation comes from studies on the survivors of the Hiroshima atom bomb. Of course the immediate effects of that bomb were horrific and dramatic. The long-term effects, which are all that need concern us, have now been studied for nearly five decades, but perhaps the effects of radiation show up even later than that. Perhaps the effects of radiation do not stay a constant risk year after year into the future but perhaps they increase with age. There is, therefore, some uncertainty about the effects which a dose of radiation at one time, say now, will produce in future years, perhaps decades away. While this element of uncertainty remains we are left with a nagging doubt that perhaps, in some way, manmade radiation is not equivalent to natural radiation and therefore maybe the comparisons which I have given you so far are invalid for reasons we cannot explain.

Quite apart from this ill-defined but fundamental point, we need to deal with our increasing scientific understanding. As new results become available, it becomes fashionable to propose different relationships, linking health risks to radiation dose. The International Commission on Radiological Protection has in fact recently issued new guidelines on this matter. They do not change our thinking in any dramatic fashion but they emphasize the fact that we are not absolutely certain about these risks.

For these risks we need to ask the following questions. We have made a comparison between the risks from natural radiation and man-made radiation and we have relied upon our present scientific understanding to produce a comparison between man-made risks and natural risks. How much is this comparison depending upon our assumptions and how much is the comparison unique and totally independent of the details of our assumptions? If we can demonstrate that the comparison is independent of our assumptions then we have done something worthwhile because we will then have demonstrated that whatever the radiation risk we postulate, it must be occurring in nature already, and must therefore be modest in size.

This mathematical proof is one I have presented elsewhere (Marshall *et al.* 1983). I will here spare you the algebra and simply explain the final result.

Whatever the uncertainty between the radiation dose received today and the death risk in the future, all reputable scientific opinion agrees that the risk is, at worst, linear in the dose received. Furthermore, if we are assessing the risks to society we want to know the average effect of that dose whether it is given in a single accident or distributed over time. For example, it is useless to say that motor cars kill some people and do not kill others, we want to know what is the probability that motor cars kill people, i.e. we want to know 'the average effect' of motor cars.

But if we want to know 'the average effect' then for natural radiation we want to average over the life of an individual and for an accident we want to average over the ages of the population exposed to the accident. But these averages must be identical because, in a single lifetime, an individual samples every age with the same probability of the age distribution of a population.

This last conclusion is a very powerful theory because it does not depend upon any assumptions about the health effects of radiation except that those effects are proportional to the radiation received according to some formula which we do not need to know. Suppose therefore we consider the effects of the dreadful Chernobyl accident. We know that, at Chernobyl, the population living nearby received on average a dose of 100 millisieverts (mSv).[1] Now the natural lifetime dose in the Ukraine is

200 mSv so, for this particular population, the effect of the accident is one half the natural lifetime effects. In other words, the effect upon that particular population will be, on average, exactly as though they were born in an area of the country where the natural lifetime dose is 300 mSv rather than 200 mSv.

Now I mentioned before that natural radioactivity is distributed throughout the continental shelf and we cannot anywhere escape it. But the distribution of natural radioactivity is not at all even. In some parts of the world, like the State of Kerala in India, or in Brazil, the radioactive level is ten times the average. Even within the UK, there is a big change, say, between the Home Counties and Aberdeen or the Home Counties and Cornwall. So a change in natural lifetime dose from 200 mSv to 300 mSv is quite routine in everyday life and has no adverse observable effects.

But we have just proved that the sudden radiation dose produced by the Chernobyl dose has exactly the same effect taking the population as a whole and we have proved that quite independently of the precise law of cause and effect between radiation and its health effects. Therefore we can be extremely confident that the effects of radiation at Chernobyl must lie well within the variations which come about because of different levels of radiation in different places.

It is for these fundamental reasons that we can remain confident that the latest scare story from Chernobyl or from Sellafield is just that, a scare story, not something that will survive careful scientific examination. Please notice we could not make this argument for the adverse environmental effects of acid rain, the ozone hole, or carbon dioxide emissions. This is an argument we can make uniquely for radioactivity and nuclear power.

Please also note that the result only applies 'on average'. We could instead ask for the effects of Chernobyl upon children exposed to radiation. We would then discover that children are affected more and adults correspondingly less than the equivalent natural radiation dose. This is because 'children' are not an 'average population'. Nevertheless, it is impossible to imagine an age-independent risk effect which is so extreme that the 'average' results are actually misleading.

I have now given you three separate discussions of nuclear power and its impact on the environment: comparative, philosophical, and mathematical. I hope that I have done something to convince you that nuclear technology does not deserve the unpopularity which it has at the present point in time.

NOTE

[1] A mSv, millisievert, is a measure of radiation weighted with the biological effect of each radiation type.

REFERENCE

Marshall, W. C., Billington, D. E., Cameron, R. F., and Curl, S. J. (1983). Big Nuclear Accidents. AERE Report No. R10532, HMSO.

Index